"*Nuclear Disaster at Fukushima Daiichi* is one of the first and most comprehensive social scientific analyses of the natural and human-made disaster that is Fukushima Daiichi. It brings together some of world's leading thinkers on science, technology and society, risk analysis, energy policy as well as indigenous Japanese scholars offering an internal critical account of the reasons, actors, dynamics and implications of this disaster. This is a major scholarly contribution to an extremely pressing and urgent issue and Hindmarsh is to be congratulated in bringing together such an impressive array of scholarship in such a short space of time."

—**John Barry**, Queens University, Belfast

"Richard Hindmarsh has added a new dimension to the global policy debate over the safety of nuclear energy. The twelve chapters in the book provide rich sources of information and conceptual agendas. The book will become a 'must' for those who want to partake in this ongoing discussion."

—**Akira Nakamura**, Meiji University, Tokyo

Nuclear Disaster at Fukushima Daiichi

Nuclear Disaster at Fukushima Daiichi is a timely and groundbreaking account of the Fukushima Daiichi nuclear meltdown amidst an earthquake and tsunami on Japan's northeast coastline on March 11, 2011. It provides riveting insights into the social and political landscape of nuclear power development in Japan, which significantly contributed to the disaster; the flawed disaster management options taken; and the political, technical and social reactions as the accident unfolded. In doing so, it critically reflects on the implications for managing future nuclear disasters, for effective and responsible regulation and good governance of controversial science and technology, or technoscience, and for the future of nuclear power itself, both in Japan and internationally.

Informed by a leading cast of international scholars in science, technology and society studies, the book is at the forefront of discussing the Fukushima Daiichi disaster at the intersection of social, environmental and energy security and good governance when such issues dominate global agendas for sustainable futures. Its powerful critique of the risks and hazards of nuclear energy alongside poor disaster management is an important counterbalance to the plans for nuclear build as being central to sustainable energy in the face of climate change, increasing extreme weather events and environmental problems, diminishing fossil fuel, peak oil and rising electricity costs.

Adding significantly to the consideration and debate of these critical issues, the book will interest academics, policy makers, energy pundits, public interest organizations, citizens and students engaged variously with Fukushima itself, disaster management, political science, environmental/energy policy and risk, public health, sociology, public participation, civil society activism, new media, sustainability, and technology governance.

Richard Hindmarsh is Associate Professor in the Griffith School of Environment and Centre for Governance and Public Policy, Griffith University, Brisbane, Australia. His field is environmental politics and policy, and science, technology and society. He is cofounder of the Asia-Pacific Science, Technology and Society Network. He has produced eight books, including *Edging Towards BioUtopia* (University of Western Australia Press, 2008) and *Genetic Suspects* (Cambridge University Press, 2010; coedited with Barbara Prainsack). Current research topics include genetically modified crops, wind and nuclear energy, good governance, community engagement, and sustainability transitions.

Routledge Studies in Science, Technology and Society

1. **Science and the Media**
   Alternative Routes in Scientific Communication
   Massimiano Bucchi

2. **Animals, Disease and Human Society**
   Human-Animal Relations and the Rise of Veterinary Medicine
   Joanna Swabe

3. **Transnational Environmental Policy**
   The Ozone Layer
   Reiner Grundmann

4. **Biology and Political Science**
   Robert H Blank and Samuel M. Hines, Jr.

5. **Technoculture and Critical Theory**
   In the Service of the Machine?
   Simon Cooper

6. **Biomedicine as Culture**
   Instrumental Practices, Technoscientific Knowledge, and New Modes of Life
   Edited by Regula Valérie Burri and Joseph Dumit

7. **Journalism, Science and Society**
   Science Communication between News and Public Relations
   Edited by Martin W. Bauer and Massimiano Bucchi

8. **Science Images and Popular Images of Science**
   Edited by Bernd Hüppauf and Peter Weingart

9. **Wind Power and Power Politics**
   International Perspectives
   Edited by Peter A. Strachan, David Lal and David Toke

10. **Global Public Health Vigilance**
    Creating a World on Alert
    Lorna Weir and Eric Mykhalovskiy

11. **Rethinking Disability**
    Bodies, Senses, and Things
    Michael Schillmeier

12. **Biometrics**
    Bodies, Technologies, Biopolitics
    Joseph Pugliese

13. **Wired and Mobilizing**
    Social Movements, New Technology, and Electoral Politics
    Victoria Carty

14. **The Politics of Bioethics**
    Alan Petersen

15. **The Culture of Science**
    How the Public Relates to Science across the Globe
    Edited by Martin W. Bauer, Rajesh Shukla and Nick Allum

16 Internet and Surveillance
The Challenges of Web 2.0 and Social Media
Edited by Christian Fuchs, Kees Boersma, Anders Albrechtslund and Marisol Sandoval

17 The Good Life in a Technological Age
Edited by Philip Brey, Adam Briggle and Edward Spence

18 The Social Life of Nanotechnology
Edited by Barbara Herr Harthorn and John W. Mohr

19 Video Surveillance and Social Control in a Comparative Perspective
Edited by Fredrika Björklund and Ola Svenonius

20 The Digital Evolution of an American Identity
C. Waite

21 Nuclear Disaster at Fukushima Daiichi
Social, Political and Environmental Issues
Edited by Richard Hindmarsh

Nuclear Disaster at Fukushima Daiichi

Social, Political and Environmental Issues

Edited by Richard Hindmarsh

NEW YORK AND LONDON

First published 2013
by Routledge
711 Third Avenue, New York, NY 10017

Simultaneously published in the UK
by Routledge
2 Park Square, Milton Park, Abingdon, Oxfordshire OX14 4RN

First issued in paperback 2014

Routledge is an imprint of the Taylor & Francis Group, an informa business

Library of Congress Cataloging-in-Publication Data

Nuclear disaster at Fukushima Daiichi : social, political and environmental issues / edited by Richard Hindmarsh. — 1st ed.
p. cm. — (Routledge studies in science, technology and society ; v.21)
1. Fukushima Nuclear Disaster, Japan, 2011. 2. Tohoku Earthquake and Tsunami, Japan, 2011. 3. Nuclear power plants—Accidents—Japan—Fukushima-ken. 4. Environmental disasters—Japan—Fukushima-ken. 5. Fukushima-ken (Japan)—Environmental conditions. I. Hindmarsh, R. A. (Richard A.)
TK136.J3N83 2013
363.17'990952117—dc23
2012045708

ISBN 13: 978-0-415-52783-5 (hbk)
ISBN 13: 978-1-138-83327-2 (pbk)

Typeset in Sabon
by Apex CoVantage, LLC

Contents

Figures

Tables

Contributors

Anders Blok is Assistant Professor in Sociology at the University of Copenhagen, and was recently a visiting fellow with the Program on Science, Technology and Society (STS) at Harvard University. He coauthored (with Torben Elgaard Jensen) a book introducing the science studies of Bruno Latour (Routledge, 2011), and publishes on science and technology studies, sociological theory and environmental politics. Current research is on sociological approaches to urban climate change interventions.

Andrew Blowers is Emeritus Professor at the Open University, United Kingdom. He specializes in the politics of nuclear energy and radioactive waste. Major publications include *Planning for a Sustainable Environment* and *The International Politics of Nuclear Waste*. Formerly a county councilor, he was a member of government advisory bodies, including the UK's Committee on Radioactive Waste Management, and was a board member of Nirex. Presently, he is co-chair of the Government/NGO Forum, and chair of a local antinuclear nongovernmental organization. In 2000, he was awarded the Order of the British Empire for services to environmental protection.

Catherine Butler is Research Fellow at Cardiff University. Her background is in sociology and her research interests include socio-environmental risk governance, social and cultural aspects of climate change, flood policy, public engagement and justice issues associated with energy systems and low carbon transitions. Current research examines public perspectives on and governance of climate change and energy systems. Recent articles include 'Morality and Climate Change: Is Leaving Your TV on Standby a Risky Behaviour?' in *Environmental Values*.

Jim Falk is Professorial Fellow at the University of Melbourne, and, amongst other things, is Director of Climate Change Research for the Association of Pacific Rim Universities World Initiative. He researches the nature, impact and management of science and technology in social context on issues such as governance, globalization, technological change and the environment, nuclear technology, arms races and militarization and information and communication technology. His latest book (with Joseph Camilleri) is *Worlds in Transition: Evolving Governance across a Stressed Planet* (Edward Elgar, 2009).

Takuji Hara is Professor of the Management of Technology, Business Administration, Kobe University, Japan (PhD, University of Edinburgh). His interests are in technological change, in particular the social shaping process of technology, in various industries including pharmaceuticals, transportation and energy. Current research is on the social shaping of technological safety. Books include *Industrial Innovation in Japan* (Routledge, 2008; coedited with Norio Kambayashi and Noboru Matsushima).

Richard Hindmarsh is Associate Professor in the Griffith School of Environment and Centre for Governance and Public Policy, Griffith University, Brisbane, Australia. His field is environmental politics and policy and science, technology and society. He is cofounder of the Asia-Pacific Science, Technology and Society Network. He has produced eight books, including *Edging Towards BioUtopia* (University of Western Australia Press, 2008), and *Genetic Suspects* (Cambridge University Press, 2010; coedited with Barbara Prainsack). Current research topics include genetically modified crops, wind and nuclear energy, good governance, community engagement, and sustainability transitions.

Kohta Juraku is Assistant Professor at Department of Humanities and Social Sciences, School of Science and Technology for Future Life, Tokyo Denki University, Japan. His research field is the sociology of energy technologies, centering on social decision-making processes on energy technologies, including the siting process of nuclear power and wind power facilities. An analysis on nuclear plant siting was published in the first issue of *East Asian STS* journal (2007). He joined the Global COE Program 'Nuclear Education and Research Initiative' at the University of Tokyo in 2008.

Denisa Kera is Assistant Professor at the National University of Singapore and Asia Research Institute, Singapore. Current research fuses science, technology and society studies and interactive media design, as well as public participation in science following the convergence of web technologies and biotechnology around (Do It Yourself) DIYbio movements, consumer genomics and various citizen science projects. She also has extensive experience as a curator of exhibitions and projects related to art, technology and science.

Shuhei Kimura is Associate Professor in Social Anthropology at Fuji Tokoha University, Japan. He studied cultural anthropology and has published in peer-reviewed journals of anthropology in Japan, including *Bunkajunruigagu, Japanese Journal of Cultural Anthropology* and *Japan Journal for Science, Technology, and Society*. Current research is on anthropological approaches to disaster recovery and preparedness in Japan and Turkey.

Ian Lowe is Emeritus Professor of Science, Technology and Society at Griffith University, an adjunct professor at two other universities, a Fellow of the Australian Academy of Technological Sciences and Engineering and President of the Australian Conservation Foundation. He has filled senior advisory roles to all levels of government, reviewed several major international reports and received many awards for his work in the areas of energy policy and environmental studies.

Atsuro Morita is Associate Professor in Anthropology at Osaka University, Japan. He has studied anthropology and indigenous engineering development in Thailand and technology transfer from Japan to Thailand. He is currently preparing an ethnographic book on indigenous engineering titled *Engineering in the Wild: Co-Development of Technology and Sociality in Thai Small-Scale Industry*. Current research is on indigenous technology development in the Thai informal manufacturing sector.

Karen Parkhill is Research Fellow in the School of Psychology at Cardiff University, United Kingdom, and has a human geography PhD. Research interests include risk perceptions, constructions of place, people's perceptions of energy systems and engagement with how energy is consumed and public perceptions of emergent technologies (e.g. geo-engineering). Recent publications include 'From the Familiar to Extraordinary: Local Resident's Perceptions of Risk when Living with Nuclear Power in the UK' in *Transactions of the Institute of British Geographers*.

Radka Peterova, MA, graduated from Charles University, Prague, in 2011. Her research focus was on citizen science, participatory sensing and urban computing. She initiated the start of an open hardware project that promotes awareness of the problems associated with air pollution. The project won a Social Innovation Award in the Czech Republic in 2012.

Nick Pidgeon is Professor of Environmental Psychology at Cardiff University, United Kingdom. His research looks at risk perception, risk communication and public engagement around environmental controversies such as nuclear power, climate change, agriculture and nanotechnologies. He is coeditor (with Roger Kasperson and Paul Slovic) of *The Social Amplification of Risk* (Cambridge University Press, 2003). Recent publications include 'The Role of Social and Decision Sciences in Communicating Uncertain Climate Risks' in *Nature Climate Change*.

Sara. B. Pritchard is Assistant Professor in the Department of Science and Technology Studies at Cornell University. She specializes in environmental history, the history of technology and their intersection, and is the author of *Confluence: The Nature of Technology and the Remaking of the Rhône* (Harvard University Press, 2011). Current research examines the circulation of hydraulic knowledge and management technologies between France and French North Africa during the colonial and post-colonial eras.

Jan Rod is a design researcher and designer from Czech Republic. He recently submitted his doctorate, School of Media Design, Keio University, Tokyo, Japan. His main area of research is urban computing and the design of technologies and artifacts, both digital and physical, for public space. In his PhD thesis, a strong focus was on describing the theoretical framework for emerging technologies and services that connect users with different scales of the space through data.

Sonja D. Schmid is Assistant Professor in Science and Technology Studies, Virginia Polytechnic Institute and State University, USA. She researches the history of technology, science and technology policy and social studies of risk, especially the history and organization of nuclear industries in the former Soviet Union and Eastern Europe, analyzing how national energy policies, technological choices and nonproliferation concerns shape each other. Her upcoming monograph is tentatively titled *Producing Power: Pre-Chernobyl Origins of the Soviet Nuclear Industry* (MIT Press, 2013).

Acknowledgments

First and foremost I thank my contributors, who represent a leading cohort of global scholars—situated in science, technology and society (STS) studies—concerned about the profound issues and implications raised by the Fukushima Daiichi nuclear disaster. I particularly thank our Japanese contributors who wrote in English as a second language, and to all contributors for their ready engagement in the rigorous editing process. I thank Ian Lowe for providing the foreword. I am grateful to my proofreader, Wendy Smith, for her final critical reading of the manuscript before its submission to Routledge. Thanks also to my wife, A'edah Abu Bakar, for all her support, especially through a difficult period in the book's production when I suffered a total ruptured Achilles tendon. I express my appreciation to Natalja Mortensen, the acquisitions editor, Political Science Research, Routledge, New York, who embraced the production of this collection, and editorial assistant at Routledge Darcy Bullock, for their support and excellent assistance. I also express my appreciation to the anonymous reviewers for Routledge and the copyeditor.

Also most helpful was the research support of the Centre for Governance and Public Policy, Griffith University, Brisbane. The research platform for my 'place-change planning' concept advanced in Chapter 4 was provided by an Australian Research Council Large/Discovery Grant: 'Meeting 2020 Targets: Effective Transitions for Renewable Energy & Beyond' (DP0986201), 2009–2011. Accordingly, Chapter 4 is a significantly modified version of ' "Liberating" Social Knowledges for Water Management, and More Broadly Environmental Management, through "Place-change Planning" ' by Richard Hindmarsh, *Local Environment* 17, no. 10 (2012): 1121–1136 (this is an electronic version of an article published in *Local Environment,* ©2012 Copyright Informa; *Local Environment* available online at http://dx.doi.org/10.1080/13549839.2012.729564, reprinted by permission of Taylor & Francis, http://www.tandfonline.com). Chapter 7 is a modified version of 'An Envirotechnical Disaster: Nature, Technology, and Politics at Fukushima' by Sara Pritchard, *Environmental History* 17 (2012): 219–243. With copyright with Sara Pritchard, Oxford Journals is gratefully acknowledged. In turn, Chapter 8 is a modified version of 'Nuclear Power after Japan: The Social Dimensions' by Catherine Butler, Karen A. Parkhill and Nicholas F. Pidgeon, *Environment: Science and Policy for Sustainable Development* 53, no. 6 (October 2011): 3–14, reprinted by permission of Taylor & Francis (http://www.tandfonline.com).

Richard Hindmarsh

Editor's Preface

By coincidence, immediately after the Fukushima Daiichi meltdown occurred on March 11, 2011, I visited South Korea, officially the Republic of Korea, which is separated from neighboring Japan to the east by the Korea Strait of about 200 kilometers (120 miles) at the closest point (discounting Tsushima Island). The two countries are thus very close and, as such, South Korea and Japan could easily be exposed to any transboundary environmental risks posed to each other, such as air and water pollution. My first port of call was in the south of this beautiful country at Gwangju (about 1,400 kilometers from the Fukushima Daiichi power plant), where I had been invited to present a keynote address at the 2011 International Symposium on Offshore Wind Energy Technology (March 16–17). My talk was called 'From Land to Sea: Place, Participation, Offshore Wind Farms, and Policy Learning'. This talk and the location of its delivery later contributed to my critical reflection on a similar problem with the siting of nuclear power plants in Japan, and the idea for this book.

The talk addressed how the siting of wind farms—as another controversial energy technology—raised many place-disruptive issues at the local community level, which, most often were not well understood or addressed by centralized energy planning systems. This failure had led to deep social conflict, typically in affected rural communities, as a rising policy problem of wind farm planning and development internationally, especially in terms of new global trends that gathered strength in the 2000s towards participatory planning in emphasizing 'localism' and 'community'. In turn, this social conflict has had a negative impact on effective renewable energy transitions, which highlights a key problem underpinning the many local social, economic and environmental issues of wind farms as the neglect by planning systems of local voices to have their say about such siting in their 'place'. Regarding the siting of nuclear power plants in Japan, I was soon to learn that poor community engagement in their siting was undoubtedly a key contributing factor to the flawed siting of the Fukushima Daichii power plant, which had 'invited' disaster, alongside many other contributing factors.

After my talk at Gwangju, five days after Fukushima had occurred, I was surprised to have still heard little about it in South Korea. Perhaps, I pondered, it was still too early for the shock waves of the disaster to make impact. Leaving Gwangju, a four-hour train ride took me to Seoul, the capital of South Korea, to the north, but still about the same distance of 1,400 kilometers to Fukushima Daiichi. At Seoul National University on March 21, 2011, I presented an invited seminar on regulatory issues on the lack of any meaningful citizen engagement in Australia in the decision making on the environmental release of genetically modified (GM) crops—another high-impact controversial technology, like wind farms and nuclear power. All inform my research interest as situated in the field of science, technology and society (STS) and environmental studies, which informs this book intimately on the nuclear disaster at Fukushima Daiichi and the associated social, political and environmental issues and implications raised (as Chapter 1 introduces). In Seoul, I started to hear more about the fallout from Fukushima from STS colleagues who were clearly disturbed about it, mentioning fears of wind carrying radioactive pollution to their shores, but who were relieved the winds at this time of the year typically blew to the east of Japan and not to the west. That night on the TV, in perhaps responding to the now high media attention, hourly government announcements began to advise citizens what to do to guard against radioactive pollution if the wind direction changed perchance. People were advised to stay indoors.

Such is the worldwide personal dread of radioactive pollution; I was then glad to be soon departing South Korea for home in Brisbane, Australia. Fukushima already had made an indelible mark on me, particularly as a citizen of a nonnuclear nation where I had not been exposed previously to the chilling fear of radioactive pollution. By coincidence the night before I was to leave South Korea, my wife rang me and pleaded for me to come home with the startling line, 'I don't want to drown alone!' This was because yet another significant natural disaster had just occurred there with the Brisbane floods of 2011. That nature was to blame for this disaster reflected the normal sensationalist media coverage of natural events impacting on human affairs, as was also made of the Fukushima Daiichi disaster, with an earthquake and subsequent tsunami the culprits. However, in both cases, once the initial shocks had subsided, the human or social contribution to these disasters became blatantly apparent, and indeed, was significant. It was obvious in both cases that poor decision making and management had invited high vulnerability to natural disaster.

In the case of the Brisbane floods, poor operating processes and decisions on the release of waters from the very dam meant to protect the city from flooding saw high volumes of water released haphazardly and at the wrong times. In the background were water authorities, paranoid about releasing storage waters when only two years previously, severe Queensland droughts had even threatened the city. In Japan, a long and cozy association between pro-nuclear government and nuclear industry interests had also led to safety being compromised, in both the operation and siting of nuclear power plants, as the analysis of this book well demonstrates. In short, the Brisbane flood and the Fukushima Daiichi nuclear disasters highlighted the fact that both disasters were caused significantly by a synergy of social (human) and natural factors. Making their impact more significant was their association with two strong fears of human death—by drowning or by radioactive pollution—apart from the many other fears, worries and impacts of social, economic and environmental and place disruption and upheaval, such as the loss of home and property. In the end, both disasters made an indelible mark on me from these experiences and reflections, not only personally but also professionally

Professionally, because the conjunction of social, technological and natural factors and the politics of place, technology, development and environment characterizing these disasters, especially the nuclear one, informed my emergent research agenda. This agenda involves the critical analysis of the decision making—particularly the role of community engagement or lack thereof—informing the siting of controversial technological assemblages that pose place-disruptive 'radical change' in unpredictable and complex terrains of coproduced social, technological and environmental systems (see Chapter 4 for detail). Thus, in intertwined contexts of the personal and professional, climate change adaption and sustainability transitions, science and technology as sociotechnical systems, and good governance of science, technology, energy and the environment, my interest in producing this book was well seeded.

This interest was furthered as the cofounder and then convener of the Asia-Pacific Science, Technology and Society Network (with over 230 STS scholars from 11 countries in the Asia-Pacific: see bit.ly/APSTSN). The APSTSN has a strong focus on generating practical insights for policy making in the Asia-Pacific region on science and technology, particularly on environmental issues. I was thus in a good position to call for essays on the Fukushima Daiichi disaster in STS context, and with Routledge as a 'STS' publisher subsequently enthusiastic. In attracting writers for the book, it was deemed essential to have 'local' participation, knowledges and embedment involved in reflecting place ownership and legitimacy of discussion and analysis. This work therefore sees a good contribution of Japanese scholars alongside regional ones, and because of the global impact of the disaster, scholars outside the immediate region were also invited to comment: not surprisingly they all come from 'nuclear nations' to contribute a deep critical, reflective scholarship on nuclear energy. The resultant cast of authors in this book and its publication only two years postdisaster thus makes it a highly useful and leading early contribution to the understanding of Fukushima Daiichi, the issues and implications raised and the suggestions made to address them in remedial, reformist and/or challenging contexts of long-term social and environmental well-being and sustainability in relation, in particular, to appropriate energy choices, safety and futures and overall to good governance of technology, society and the environment.

Richard Hindmarsh

Foreword

Would a rational person even consider nuclear power now? At the turn of the century, nuclear power was seen as a failed technology. Originally promoted to the public and politicians as cheap, clean and safe, it was widely recognized by 2000 as expensive, dirty and dangerous. Cancellations and deferments had been out-numbering new construction projects for decades. History may well show the Fukushima Daiichi accident as the last nail in the coffin of the nuclear industry, fatally exposing the delusion that we could safely harness nuclear fission to deliver large-scale energy. Future generations may well see it as having been a crucial wake-up call, spurring the development of genuine clean energy systems.

The nuclear power industry is a classic study in the social, political and environmental issues involved in the development and application of new technology. Public science under wartime pressure refined understanding of nuclear fission, enabling development of the weapons that effectively ended World War II. The technology of weapons production spread, and the first 'nuclear power station' in the United Kingdom essentially generated electricity with the waste heat resulting from the production of fissile material for British nuclear weapons. There followed a brief period of extreme enthusiasm for 'atomic energy', extending from power stations to nuclear submarines and even plans for nuclear-powered aircraft. The public was promised energy that would be so cheap its use would not even be metered and charged. The industry's problems steadily grew, as construction delays and cost overruns undermined that optimistic promise. By 1980, it was clear that nuclear energy was not cost-competitive with fossil fuel generation, so planned construction projects were being delayed or shelved. The Three Mile Island accident in 1979 was a critical blow to the US nuclear industry, but worse was to come. When the Chernobyl accident scattered radioactive debris over large areas of Europe, observers thought that the industry would never recover. Pronuclear enthusiasts attempted to persuade the public that Chernobyl was a consequence of design or operation problems that were peculiar to Eastern Europe, but there is no convincing evidence that explanation dispelled concerns about the technology. By 2000, the industry appeared to be in terminal decline.

There followed an ingenious campaign, begun in the United Kingdom and adopted enthusiastically in other countries, to reframe nuclear energy in the minds of decision makers and the public. The industry abandoned its clearly ineffective approach of portraying nuclear power as the clean, technically advanced energy supply for the future, since this line was clearly failing to impress communities worried about accidents and other risks. Instead, the nuclear industry—in embracing the arguments of environmentalists and climate scientists, publicly accepted that the science demanded a concerted response to climate change—proposed nuclear reactors as the 'only proven low-carbon electricity supply technology'.

This claim by the proponents of nuclear power, that it is the only way to produce large-scale electricity without the carbon dioxide emissions that are changing the global climate, did indeed reframe the debate. Enthusiasts even began talking of a 'nuclear power renaissance', although critics pointed out that this was really only a revival of talk about nuclear power and proposals for power stations, rather than concrete decisions to build new reactors. The claim did, however, raise a serious issue that deserved analysis and public debate. The science of climate change shows that the burning of fossil fuels is posing a serious risk to human civilization. A rational response demands a rapid move away from fossil fuel use in general and coal-fired electricity in particular. If nuclear power were indeed the *only* effective way of slowing climate change, it might be possible to make a rational case for building new reactors.

We would then, however, have to address the two fundamental problems that the Fox Inquiry identified in its report to the Australian government in 1976: radioactive waste and weapons proliferation. If we were to continue using or even expand nuclear energy, we would need to put a huge effort into managing its waste. For 50 years, the industry has essentially been stockpiling high-level radioactive waste and assuring the community that a long-term solution would be found. This is a formidable challenge, since some components of the waste need to be isolated from natural systems for hundreds of thousands of years: much longer than any human civilization has ever endured. While this is a massive challenge to our social institutions, waste management is primarily a technical problem that might eventually be solved, at least in principle, given enough resources and political commitment. There are other social issues involved besides the need for durable safety systems that will survive the end of our current civilization. Even in communities that strongly support mining and export of uranium, there is invariably determined opposition to proposals for storing low-level waste. In most countries with long-standing nuclear power programs, the existence of determined local opposition has proved a barrier to establishing radioactive waste facilities.

Even if the waste problem were solved, we would still have to worry about proliferation of nuclear weapons, as this is a social and political problem with no apparent prospect of solution. While it was possible for a few decades to put our faith in the Nuclear Non-Proliferation Treaty, we now have to recognize that it has been a failure. The nations having nuclear weapons when the treaty was negotiated have not disarmed; in some cases, they are still actively developing their nuclear arsenals. In the absence of an effective disarmament regime, several other nations have developed their own nuclear weapons: India, Pakistan, Israel, probably North Korea. Any nation with the technical capacity to use nuclear energy also has the wherewithal to develop weapons. The recent example of North Korea shows that there are no realistic sanctions against those who take the next step; once a leader has nuclear weapons, no military challenge is feasible.

The more people use nuclear technology, the greater is the risk of fissile material being diverted for weapons. The problem is likely to get steadily worse until we actually see nuclear weapons again used in armed conflict. That is a truly awful prospect, as modern bombs are literally thousands of times more powerful than those used in 1945. The problem is not confined to national governments. The former head of the International Atomic Energy Agency, Dr Mohammed El Baradei, told the UN that he faced the impossible task of regulating hundreds of nuclear installations with a budget comparable to that of a city police force. His agency documented literally hundreds of examples of attempts to divert fissile material for improper purposes. There is a real risk of rogue elements or terrorists having either full-scale nuclear weapons or the capacity to detonate a 'dirty bomb' that could make an entire city uninhabitable.

So embracing nuclear energy on an expanded scale would truly be a Faustian bargain. Fortunately, we do not need to face that terrible dilemma. There are other, much better ways of cleaning up our energy supply system. The nuclear argument is a dangerous distraction that could direct resources and technical capacity away from more sensible responses.

Nuclear power is also certainly not a fast enough response to climate change. A strongly pronuclear advisory group set up by the Australian government to consider the issue concluded that it would take 10 to 15 years to build *one* nuclear reactor in that country. Their proposed crash program of 25 reactors by 2050 would only slow the *growth* in Australia's greenhouse gas production, not achieve the reduction that is needed. While the delay is naturally greater in countries that do not have an existing nuclear power industry, it remains generally true that it would take much longer to construct a given amount of nuclear capacity than to bring on line equivalent amounts of wind or solar power.

Secondly, it is too expensive. Again, even pronuclear expert groups usually find that there needs to be both a carbon price and other government subsidies to make nuclear look competitive with alternatives such as wind power on good sites. Even solar energy does not look much more expensive than nuclear power, taking into account the real cost and timescale of construction rather than the claims of the industry. Optimists tell us to wait for a promised new generation of reactors that they believe could be cheaper and safer, but even if we believed assurances from authorities who have been consistently wrong for decades, we cannot afford to delay tackling climate change in the hope that new reactors might overcome the problems of existing installations.

While those who support nuclear energy say that new nuclear power stations would not have the technical limitations of Three Mile Island or Chernobyl, or be sited as dangerously as Fukushima Daiichi, there will always be some risk of accidents. An accident in a nuclear power station is a much more serious risk than an accident in any form of renewable energy supply. As the Fukushima Daiichi accident demonstrated, the release of significant amounts of fissile materials has a wide range of impacts over a range of temporal and geographical scales. By comparison, the clean energy response is quicker, less expensive and less dangerous. There is no risk from terrorists stealing solar panels or wind turbine blades. We know how to decommission wind turbines and solar panels at the end of their life, at little cost and with no risk to the community.

This book is an excellent compilation of analysis of the 'fallout' from Fukushima: social, economic, political, environmental and technical. It provides a rich diversity of perspectives on the issue and a sobering reflection on the long-term implications of the accident. It has been wisely said that those who do not learn the lessons of history are condemned to repeat the problems of the past. We cannot turn back time and prevent the Fukushima Daiichi accident, but we owe it to future generations to learn from it. It may well prove to have been the stimulus to develop the clean energy systems and efficient use practices that we will need if civilization is to survive. It also provides a framework for reevaluation of the role of technical experts in providing policy advice about the uses of science and technology. As this case study demonstrates, that reevaluation is urgently needed.

More generally, the case study also demonstrates the efficacy of the STS approach. The broad area of science, technology and society studies provides a framework for analysis of these complex issues. Unless we recognize the social and political context in which the decisions about science and technology are made, we will perpetuate the culture of expertise that has allowed a succession of nuclear accidents. Similar invalid assurances about the capacity of technical experts to foresee all eventualities have underpinned the failure of most governments to regulate other advanced technologies, such as nanotechnology and the release of genetically modified organisms. Ridiculous proposals for geo-engineering projects to slow climate change need to be viewed through the same lens. An informed and critical community is the only effective defense.

Ian Lowe

1 Nuclear Disaster at Fukushima Daiichi

Introducing the Terrain

Richard Hindmarsh

The tragic effects and implications of the Fukushima Daiichi nuclear power plant disaster—which many refer to as Fukushima—will continue to reverberate over the coming years and decades. The implications assume a new regional and global scope that build on but go beyond those of prior nuclear meltdowns at Three Mile Island (1979) and Chernobyl (1986). As such, they renew and reinvigorate past memories and impacts of trauma, terror, stigmatization and survival, and create new ones in time and space for Japan, its immediate region and globally. Social, political and environmental issues and implications build on prior ones of nuclear power plant safety and siting, radiation pollution and health, and global energy use and choices. These are found in contexts of risk, hazard and trust; disaster management; and science, technology, environment and good governance—as informed by principles of openness, participation, accountability, effectiveness and coherence (Commission of the European Communities 2001); and new ones of climate change mitigation and adaptation, escalating global environmental problems and associated 'clean' energy choices and sustainability transitions.

In this terrain, the interest and purpose of this book—*Nuclear Disaster at Fukushima Daiichi*—is to investigate the social, political and environmental reasons for and issues and implications of the disaster in the public interest, in contribution to knowledge, and in the broader umbrella of social and environmental responsibility and justice to the victims of the disaster and to future generations; more broadly, to the long-term sustainability of interrelated social, technological and environmental systems. In this case, as mediated, negotiated or threatened by the nature of nuclear power as a 'megatechnology', as what occurred at Fukushima Daiichi well demonstrates. In certain circumstances, megatechnologies feature uncertain and unpredictable and multiple complex interactions and consequences that cannot be adequately tested in laboratories or by way of computer simulations beforehand (Beck 1995: 20). 'Rather, their unanticipated consequences can *only* be discovered *after* they are implemented' (Unger 2001: 282).

In investigating the background and lead-up to the disaster and its immediate aftermath, *Nuclear Disaster at Fukushima Daiichi* is positioned at the forefront of undoubtedly a long interrogation of 3/11 (as the disaster is also referred to); just as Three Mile Island and Chernobyl have long been interrogated and will continue to be so, despite the attempts of the global nuclear industry to overcome the 'Chernobyl syndrome' (Schneider et al. 2011. That is the more difficult to achieve also, when the *Guardian* ("Nuclear Power Plant Accidents" 2011), in reporting Fukushima, pointed up '33 serious incidents and accidents at nuclear power stations since the first recorded one in 1952 at Chalk River in Ontario, Canada'.

As such, and in being published only two years after the disaster, this book provides an important early benchmark for ongoing reflection, social and policy learning and change with regard to the failures and hazards of nuclear power in Japan and globally. It critically reflects on and constructively addresses the crisis as one of the most important events of the 21st century—for example, as an important turning point for reconsidering energy options for a world increasingly facing a future of limited energy options; of how to better deal with megatechnological risk, hazard and disaster; and of the need for energy security to be well located in good governance and long-term social and environmental sustainability.

The complexity of these posed issues and contexts, together with the broad and ongoing impact of the disaster, is well addressed by the book's multi-author cast from Japan, countries in the immediate region and more distant ones. Such a cast can best answer the inquiries of a worldwide community grappling to understand why and how the disaster occurred and what can be done in relation to its implications. The interest and concern of the global community was shown in 73,700,000 Google hits in little more than four months in the disaster's aftermath for the search term 'Fukushima' (Friedman 2012: 55).

A key implication posed by the nuclear disaster, which frames the analysis of this book, is to understand and interpret it as one caused by an unusual combination of social and natural factors. This combination reflects broader studies of technological disasters (e.g. Gramling and Krogman 1997), which would situate Fukushima Daiichi at the conjunction of a *chronic technological disaster* and a *natural disaster*, where the latter was not in play at Three Mile Island or at Chernobyl. A chronic technological disaster according to Gramling and Krogman (1997: 42) is

> predicated on mitigated, or not, by deliberate human decisions and resulting policies or lack thereof and are defined by the interplay of the various stakeholders involved. For a chronic technological disaster to occur, decisions had to be made to allow the potentially dangerous activity to go forth, or at a minimum not to oppose it. Once the initial decisions are made, additional decisions must follow concerning what safeguards, if any, should be put into effect to prevent the potential from being realized. Additional polices, or lack thereof, determine what response is possible in the event of an incident, and to a large extent

> what the impact of the precipitating incidents) will be on ecosystems, human populations and communities. Other decisions and policies determine to what extent the impacts can be mitigated and to what extent communities can recover from these effects.

Such interplay reinforces the 'perspective that chronic technological disasters should be approached from a process rather than event perspective . . .' (Gramling and Krogman 1997: 42). Examples given of such disasters include Love Canal (the New York urban toxic waste disaster in the mid-1970s), Chernobyl (1986) and the *Exxon Valdez* Alaskan oil spill (1989). These disasters all demonstrated loss of control 'over an activity that was believed to be controllable'. Other features of a chronic technological disaster include long-term and chronic levels of stress as a general outcome of significant alteration of the relationship between a community and its built, modified and biophysical environments; social disruptions to families' future expectations; and loss in belief of institutional legitimacy (Gramling and Krogman 1997: 42–44, 50). Such processes and features characterize the Fukushima Daiichi disaster as the contributors to this book have well found. We can then posit the Fukushima Daiichi disaster represents a *new type* of major nuclear disaster, which is found at the intersection of a chronic technological disaster and a natural disaster.

The *natural* at Fukushima represented a magnitude 9.0 reverse fault megathrust earthquake—the 2011 Tōhoku earthquake or Great East Japan Earthquake—occurring some 100 kilometers off the Pacific coast of Tōhoku. It was the biggest earthquake ever to affect Japan. It triggered huge tsunami waves that reached heights of up to 40 meters (130 feet). The tsunami hit the east coast of Japan including the Fukushima Prefecture and swept away entire towns, many residential areas and fishing ports, and severely damaged and destroyed industrial and commercial zones (Matanle 2011: 823). The double natural disasters resulted in more than 15,000 deaths.

The tsunami also 'breached the protective walls at the Fukushima Daiichi nuclear power plant located in Okuma and Futaba Towns, and knocked out the mains electricity supply and backup generators that [supplied] the six reactors' cooling systems' (Matanle 2011: 825–826). This breach of the power plant's defenses, in turn, directly contributed to the meltdowns and explosions of reactors 1, 2 and 3, and severe damage to reactor 4 and containment systems, which led to 'the uncontrolled leak of radioactive materials beyond the vicinity of the plant' (Matanle 2011: 826). Subsequently, the 'leak' became so pervasive that it was rated a level 7 'major accident' on the International Nuclear and Radiological Event Scale (INES), as was Chernobyl. This level, the highest on the scale, describes an event involving a 'major release of radioactive material with widespread health and environmental effects requiring implementation of planned and extended countermeasures' (IAEA 2011).

In turn, the *chronic* at Fukushima represented many human actions evident in being causative of what happened. Perhaps most *visible* were the immediate actions and responses of the Japanese Government and the Tokyo Electric Power Company (TEPCO), which ran the Fukushima Daiichi nuclear power plant, to lose control of, or make worse, the immediate situation, both inside and outside the crumbling plant after the tsunami impact. More *invisible* was an enveloping background terrain of sociopolitical endeavors involving policies, practices and actions, or lack thereof, which over decades laid the policy fabric for the disaster to occur or, again, to make it worse than it otherwise might have been.

Within the overall context of nuclear power advanced by Japan as a key energy option for the full electrification of Japan's post–World War II reconstruction and development, a key contributing factor to the disaster that the contributors of this book identify was so-called 'agency capture': 'where regulatory agencies for a variety of reasons come to hold the perspective of the industry or interests that they are established to regulate' (Gramling and Krogman 1997: 45). Subsequently, this contributed—as reinforced by the findings of this book—to the safety of nuclear power being compromised to progress it as rapidly and as unhampered as possible. Thereafter, highly misleading public representations of the safety of the aging fleet of Japan's nuclear power plants were made to retain public support and trust despite weak regulatory and flawed siting policies; flawed power plant design and regulatory assessment to withstand worst case scenarios of earthquakes and tsunamis; and a lack of transparency and accountability in decision making, including lack of public participation about hazardous technological development and siting, made more notable by highly unstable seismic conditions of siting in Japan. Overall, the chronic technological disaster of Fukushima was informed by a strategic and far-reaching 'sociopolitical terrain' to advance nuclear power development it seems at all costs. It is within such a terrain that the disaster appears most situated or determined, rather than a natural 'landscape', which instead appears complementary.

In this book a 'cartographic' approach has been adopted to map out this sociopolitical terrain including the impacts, issues and implications the disaster informed. In some ways, this cartographic approach follows the approach of French sociologist of science and anthropologist Bruno Latour (1987) called *science in action*, which involves following scientists and engineers through society to best understand the practice of what they do in an integrated sense. Thus, in this book, through the diverse but coherent perspectives of its contributors, the integrated result is what I refer to as a *policy in action* determination of why and how the Fukushima nuclear disaster occurred and the implications of that, which then provides insights of how to address the social causes of the disaster. However, of course, we cannot claim to include everything, such is the complexity of the policy in action terrain and of the actual disaster itself (see also Nakamura and Kikuchi 2011). Instead, key areas involved and impacted on are identified and addressed to provide a revealing social and sometimes technical (e.g. Falk, and Pritchard, in this book) account of the disaster at an early stage of critical reflection on it.

Regarding the sociopolitical, for example, pronuclear policy and industry interests *in action* are mapped over time to reveal a deployment of a discursive suite of maneuvers, practices and political technologies aiming to negate, marginalize or overcome the concerns, opposition or resistance of others to nuclear power and its development and ongoing expansion in Japan (following, e.g. Andrée 2002, Foucault 1990, Hindess 1982: 498, Hindmarsh 2008: 12–13). The deployment of these actions, as indicated earlier, involved a number of practices aimed at nuclear power development, safety, regulation, siting and civic involvement (e.g. Funabashi and Kitazawa 2012, Nakamura and Kikuchi 2011, Moe 2012). One clear example of a highly questionable or weak regulatory action, also reinforcing the notion of agency capture, was with regard to Fukushima Daiichi as an aging plant near the end of its lifespan with out-of-date safety systems, which Moe (2012: 270) called 'regulatory fraud' in again referring to 'collusion between the [nuclear power or electricity] utilities and the regulator . . .'.

However, before elaborating more on this policy-in-action terrain in a brief overview of the chapters, a snapshot is provided of this terrain and the social, political and environmental impacts and implications of the Fukushima Daiichi disaster, as an unfolding cataclysmic event (see also Jasanoff 1994), to better set the scene. Second, the specific objectives of the book—as informed by this terrain and broader global developments concerning nuclear power—are given. Third, a summary is given of the field of investigative inquiry of science, technology and society (STS) studies, which informs the contributors' analysis in addressing the objectives as they see fit. Finally, an overview of the key areas or themes of the disaster as identified and addressed by the book's contributors is given.

FUKUSHIMA DAIICHI AS AN UNFOLDING CATACLYSMIC EVENT: A SNAPSHOT

With 'the uncontrolled leak of radioactive materials beyond the vicinity of the plant' following the Fukushima Daiichi reactor meltdowns and damage to protective buildings (Matanle 2011: 826), a core health issue of trauma about nuclear energy confronted the Japanese people. Harmon (2012: 1) made the succinct point: 'The prospect of invisible radioactive material contaminating the air and ground [was] terrifying—especially for a country that experienced two nuclear bomb attacks in 1945'. Undoubtedly, this had some impact postdisaster on a clear majority of Japanese citizens reversing earlier positions of support for nuclear power to one of a gradual phase-out. This position has not lessened since the disaster as developments and ongoing revelations continue to shock the Japanese people and the global community.

For example, standard radiation dose limits for Japanese citizens were increased soon after the disaster, apparently to stymie evacuation from some places that were relatively, but not immediately, close to the accident site. Suggestions as to why that occurred include the financial cost would be too great to evacuate so many, as well as the social chaos caused by such upheaval. This includes places still considered by many authorities worldwide to be too close to Fukushima Daiichi—like Fukushima City, a city of some 300,000 people about 70 kilometers (45 miles) away—as the investigative documentary *Fukushima, Never Again* focused upon (also McCurry 2011, Normile 2011). Perhaps also with the discovery that radiation-contaminated beef was found dispersed and eaten widely in Japan post-disaster; as well as more than 40% of fish caught close to the Fukushima Prefecture being considered unfit for humans to eat under Japanese regulations a year postdisaster (Amos 2012).

Negative polls on nuclear power have also increased worldwide, bolstered by the continuing revelations of the spread of radioactive material from Fukushima Daiichi. For example, in early 2012 radioactive material from the nuclear disaster site was found in tiny sea creatures and ocean water some 300 kilometers (186 miles) off the coast of Japan, revealing the extent of the release: 'In some places, researchers from the Woods Hole Oceanographic Institution . . . discovered cesium radiation hundreds to thousands of times higher than would be expected naturally' (Emspak 2012). In October 2102, *Science* Magazine reported that 80% of the radioactivity from Fukushima Daiichi 'was either blown offshore or directly discharged from waters used to cool the nuclear power plants' (Buesseler 2012: 480; see also Pritchard in this book: 123–124), which immediately posed ongoing health issues for the seafood-loving Japanese (Biello 2011).

Indeed, although people oppose nuclear power for a variety of reasons that well mark its cataclysmic status when disaster strikes a primary reason is dread of radiation pollution (Freudenburg 1997). This was well indicated with 22,400,000 Google hits in little more than four months in the disaster's aftermath for the search term 'Fukushima and radiation' (Friedman 2012: 55). Also, pre-Internet, in the two weeks following Chernobyl in five US newspapers and on three major TV networks, 46% of the 394 articles and 60% of the 43 newscasts included radiation information (Friedman 2012). Other prominent and associated reasons include fear of place-associated stigmas when nuclear facilities are sited locally in the close proximity of a residential community and stress about potential contamination from living close to a nuclear facility, especially exposure of children to radiation and resultant health impacts (Freudenburg 1997), as Chernobyl made quite clear.

As such, with post-Fukushima radiation data releases by the Japanese Government unreliable, inadequate and far from prompt in the immediate aftermath of the disaster (Brumfiel 2011, Nature 2012)—alongside strong public resentment about mismanagement of nuclear safety and the disaster—a key focus of hostile Japanese publics opposed to any renewal of nuclear power in Japan was and remains the protection of children. New nongovernmental organizations and community groups like Mothers Rise against Nuclear Power

(Kakuchi 2011) and the Fukushima Network for Saving Children from Radiation remain furious that government guidelines have allowed 'some [34,000] schoolchildren to be exposed to radiation doses that are more than 20 times the previously permissible levels' (Tabuchi 2011, also Normile 2011, Falk in this book: 161). By association, the strengthening opposition postdisaster saw the launch of Japan's first green political party, Greens Japan (McCurry 2012), with a central plank being the abolition of nuclear power plants.

Not surprisingly, the Fukushima Daiichi disaster has also catalyzed a global epistemic shift already challenging widespread global moves to a nuclear 'renaissance' (Blowers 2010), or 'new build'. The push for a nuclear renaissance, which began in the early 2000s, reframes nuclear power as a low carbon alternative to fossil fuels as part of addressing rising climate change and energy security issues (Bickerstaff et al. 2008, see also the Foreword in this book), but implicitly a prime aim has been to secure renewed government spending in nuclear power (Corner et al. 2011). The benefit claim presaged is of enabling high power usage to continue whilst working toward low-carbon emission targets, with the premise that nuclear power amounts to 'clean air energy' (Duffy 2011: 670). Nevertheless, a post-Fukushima GlobeScan (June–September 2011) survey of public opinion in 23 countries found that claim unconvincing with public opinion against nuclear plants rising significantly since an earlier poll in 2005. In countries with nuclear reactors, 39% said no further reactors should be built, and a further 30% said that operating plants should also be closed as soon as possible.

This global shift against nuclear power post-Fukushima aligns with the finding of Parkhill et al. (2010) that nuclear accidents and incidents at distant locations lead to renewed anxiety and concern about nuclear power facilities everywhere, which further stigmatizes nuclear power as an energy technology (Flynn 2003). This is also shown where many countries have reconsidered their energy agendas and continue to do so since Fukushima (Bradford 2012). Germany was at the forefront in announcing the shutdown of its nuclear plants by 2022 in favor of renewable energy, apparently catalyzed by 'a popular revolt against nuclear new build . . .'. This 'revolt' saw the ruling conservative party immediately losing office to the antinuclear Greens and Social Democrats in the March 27, 2011, Baden-Württemberg state election, with 200,000 protestors marching on the eve of the election in Hamburg, Berlin, Cologne and Munich demanding closure of all 17 of Germany's nuclear reactors (Stratton 2011).

Unlike Three Mile Island and Chernobyl, Fukushima also posed highly threatening transboundary implications for countries in close proximity, including China, South Korea and Taiwan (e.g. Chan and Chen 2011, Hong 2011), which heighten regional implications of environmental, health and social risk. Posed is a precautionary need for partnership in regional governance approaches on facility siting of nuclear power plants and the substance of future energy choices and mixes. Such suggestion is reinforced by the Asia-Pacific region as a seismic zone of high to extreme crustal instability (Connor

2011), with Japanese (and nearby Taiwanese) reactors placed within the top risk group of very high seismic hazard (Cochran and McKinzie 2012, Tamman et al. 2011).

In turn, at the business end of critical comment was a study by Swiss-based investment bank UBS (2011: 1), which argued that the Fukushima disaster posed

> the most serious ever for the credibility of nuclear power. Chernobyl affected one reactor in a totalitarian state with no safety culture. At Fukushima, four reactors have been out of control for weeks—casting doubt on whether even an advanced economy can master nuclear safety.

To redress the situation, the UBS study suggested that 'some plants [would need] to be "sacrificed" to restore public confidence', with old plants close to seismically active areas or borders 'at particular risk' (UBS 2011: 1). However, the assumption that advanced economies could 'master safety' in the first place was in doubt when the International Atomic Energy Agency (IAEA) found that Japan had 'underestimated the danger of tsunamis and failed to prepare adequate backup systems at the Fukushima Daiichi nuclear plant' (see also Butler et al. in this book: 142). This finding (which reinforces findings in this book), in turn, 'seemed to report a widely held criticism in Japan that collusive ties between regulators and industry [had] led to weak oversight and a failure to ensure adequate safety levels at the plant' (Frackler 2011, also Funabashi and Kitazawa 2012, Moe 2012, Nakamura and Kikuchi 2011).

A keen political implication of the Fukushima Daiichi disaster is thus the need for a renewed focus on modifying nuclear regulatory styles to demonstrate a more responsible and transparent civic style reflecting one of good governance. The need for global regulatory reform including Japan was reinforced by Andre-Claude Lacoste, the outgoing head of the French Autorite de Surete Nucleaire (with 58 reactors under its purview) in a media interview in October 2012 (Patel 2012):

> Regulators in some countries, which [Lacoste] declined to name, lack enough independence from industry and government to be able to expose nuclear safety shortfalls, Lacoste said. In some small nations, atomic experts regularly rotate between jobs at safety authorities, operators and research organizations, creating conflicts of interest and a deep-seated reluctance to raise the alarm about potentially dangerous situations. 'In some societies solidarity between people is more important than transparency and this means that errors aren't questioned and are hard to correct. . . . There are countries where transparency isn't a virtue.' . . . Lacoste headed a peer review of Japan's nuclear regulatory system in 2007, which he said 'explicitly' warned the country of shortcomings and recommended creating an independent watchdog that would separate safety oversight from the government ministry that

> promoted atomic power. 'The Japanese situation wasn't good,' according to Lacoste, who said he was informed that the Japanese government had rejected his recommendations at the end of 2010. He has since called Fukushima a 'collective failure'.

A similar assessment was made by a third-party 12-member inquiry panel appointed by the Japanese government on May 24, 2011 (Matanle 2011: 841), 10 weeks postdisaster. In the pressure cooker of intense public and international outcry, the panel was set up to investigate the disaster and to determine what to do about it. Just over a year later, in late July 2012, the panel 'delivered a damning assessment of regulators and the station's operator' to a packed media (Maeda 2012: 841), on which a *Reuters* journalist reported:

> The panel suggested post-Fukushima safety steps taken at other nuclear plants may not be enough to cope with a big, complex catastrophe caused by both human error and natural causes in a 'disaster-prone nation' like Japan, which suffers from earthquakes, tsunami, floods and volcanoes. . . . But the inquiry stopped short of accusing the regulators and Tepco of collusion, a charge included in a strongly-worded report by the parliamentary panel earlier in July. 'The Fukushima crisis occurred because people didn't take the impact of natural disasters so seriously,' [panel chair Professor Yotaro] Hatamura told a news conference . . . (Maeda 2012: 841)

In the immediate aftermath of Fukushima, the global nuclear industry had also highlighted natural causes rather than those of human endeavor and error. As such, stress checks were quickly performed on many of the 436 nuclear power plants operating in 31 countries (Suzuki 2011). In Japan, all nuclear reactors were closed down for a stronger safety assessment than elsewhere. As well as to ensure adequate preventative measures were in place (Jorant 2011), this policy response aimed to restore public confidence about nuclear safety. Nonetheless, these measures had little impact on the announced nuclear power phaseouts of Germany (for 2022), Switzerland (2025) or Belgium (2034), the rejection by referendum of plans to revive Italy's nuclear program and growing numbers of dissenting Japanese citizens.

Indeed, in Japan, public anxiety about nuclear power was at an all-time high, informed by distrust and suspicion of government and industry about the exact reasons for Fukushima, apart from the obvious contribution of natural ones. Questions flooded the media and the civic sphere, including: Why had the official release of radiation data to inform citizens been so slow? (The real-time spread of radioactivity had been withheld from the public for 12 days after the crisis began [Nature 2012].) Why was radiation data unreliable and often without context for effective risk communication? Why was the evacuation of communities around the disaster site delayed? And, why was there a communication gap between the government and

TEPCO (Funabashi and Kitazawa 2012)? Such unanswered questions deepened perceptions of nuclear power, the government and TEPCO as untrustworthy and fueled the preferences for nonnuclear. They also prompted active use by citizens of the Internet, websites, blogs and social media to network with each other and try and work out the answers in the absence of reliable, prompt and coherent official data that could also be easily understood by lay citizens (see also Morita et al. and Kera et al. in this book). As Friedman (2012: 56) commented,

> these sources yielded a deluge of Fukushima information, and they have changed the definition of mass media in many ways. . . . Hundreds of Twitter conversations appeared under a variety of hashtags—such as #fukushima, #nuclear, and #meltdown—with people keeping up to date on events and where to find articles to read or videos to watch. Anyone who wanted a timeline for Fukushima events could also turn to Wikipedia, which compiled a day-by-account, including radiation readings.

In addition, in deepening and further reflecting contemporary trends to science democratization movements and science citizenship, 'during the early days of the accident, when the Tokyo Electric Power Company and the Japanese government held news briefings to provide minimal and somewhat optimistic information, their reports were quickly interpreted, supplemented, and contradicted online by scientists, government personnel, nuclear industry or anti-nuclear sources, and private individuals' (Friedman 2012: 56).

Consequently, the Japanese government became increasingly pressured by the situation as a humanitarian and electoral issue, by the increasing public preferences for denuclearization of energy, and by the increasing public protest and distrust of the government, as well as by high international interest and concern. On September 16, 2012 (16 months postdisaster), in deep damage control, and on the recommendations of its inquiry panel, the Japanese government launched a 'new regulatory body'—the Nuclear Regulation Authority. Journalist Yamaguchi (2012) remarked, 'to replace the agency whose lack of independence from the nuclear industry had been widely blamed for contributing to last year's disaster'. More dramatically, the government announced its intentions to support the panel's recommendations to phase out nuclear power by 2040, and develop a 'greater reliance on renewable energy, more [electricity] conservation and sustainable use of fossil fuels' (Yamaguchi 2012, also Soble 2012).

Nevertheless, within a short four days, under intense pressure from pronuclear business groups and communities—who argued that abandoning nuclear power would overly damage the economy—the government 'stopped short' of formally adopting the phaseout goal. Instead, as informed by the deliberations of its ministerial-level Energy and Environment Council (Onose 2012: 1), the government announced it would 'consider' these recommendations and discuss with society its crafting of a post-Fukushima energy policy that would feature 'flexibility, based on tireless verification and reexamination' for such a changing landscape.

To nuclear critics this sudden change of events posed yet another government compromise to accommodate the nuclear industry even as government ministers said they 'intended to use the [phase-out] goal as a reference point' (Tabuchi 2012: 1). Business interests were also hopeful that the center-right Liberal Democratic Party would eventually regain power because of its reluctance to 'eliminate' nuclear power (Japan Times 2012a). At the time of writing (in early November 2012), these developments marked where the politics of nuclear energy in Japan lay, with construction of a nuclear plant in Aomori Prefecture resumed, 'becoming the first utility to so do since the disaster . . .' (Japan Times 2012b). As the *Japan Times* also reported: 'The government is allowing utilities to finish building reactors that have already been approved . . . [which] is controversial, seeming to contradict another government plan to phase out nuclear power generation . . .'. In addition, in breaking news as this book went to press, the *Japan Daily Press* on November 2, 2012, reported that four members of the new independent regulatory authority had 'been discovered as accepting large amount[s] of money from those that benefit from the promotion of nuclear energy' (Westlake 2012). This revelation occurred when members were asked by the authority to voluntarily 'admit any such payments, in an effort to be transparent'. The obvious question posed was, 'can the Japanese public really trust an organization that was established on the principles of being independent from *both* the government and nuclear industry when its members have directly been give money from the latter?' (Westlake 2012).

Informed by that disturbing snapshot of the cataclysmic status of the Fukushima Daiichi nuclear disaster and its policy in action terrain, and the profound social, political and environmental issues and implications raised, what are the specific investigative objectives of the book and field of inquiry?

OBJECTIVES AND FIELD OF INVESTIGATIVE INQUIRY

Objectives

Seven objectives of the book variously inform the contributors' analysis and chapters. *First*, to critically reflect upon the disaster's social, health, environmental, technological, and political and policy issues—from a diversity of perspectives informed by the field of science, technology and society studies (as discussed later in this chapter), from Japan, the immediate region and globally. *Second*, to facilitate wide dialogue to better understand and respond to the disaster, for knowledge production, good governance and social (e.g. attitudinal) and policy learning, in Japan and globally.

Third, to analyze why and how the disaster occurred, as well as the limited emergency responses in the aftermath, which pose implications and policy lessons for what might be done to avert similar disasters in the future or to better manage them, furthered by the looming threat of increasing natural disasters as a result of climate change in addition to the implications of siting nuclear plants in areas of high seismic activity. *Fourth*, to question the safety standards of nuclear power stations and their siting in the context of Fukushima as the third major nuclear power disaster worldwide, in following Three Mile Island and Chernobyl over two decades earlier and with many advances in nuclear safety since then, and industry claims of strengthened safety.

Fifth, to identify and discuss important new sociopolitical phenomenon for policy development and good governance in relation to enhanced communication between governments and publics on social, political and environmental issues, here, those of disaster management. With regard to Fukushima, this related to new media and social networking practices of citizens, which arose during the course of the disaster to deal better with a radiation risk communication crisis because of limited, sporadic, unreliable and inaccessible radiation data broadcast by the official response team. *Sixth*, to reflect on the possibility that Fukushima represents an emerging end of the 'nuclear dream' as earlier forecast with Chernobyl, for global energy supply, as evident in many worldwide reactions to the disaster and preferences expressed for renewable energy. *Seventh*, to address a notable policy gap in environmental governance and knowledge about nuclear power siting in Japan and globally, according to notions of good governance and long-term sustainability for effective energy transitions in contexts of social and environmental risk and climate change and adaptation; which, of course, would inherently include the safety and best choice of energy systems for both humans and nonhumans.

Field of Investigative Inquiry

In addressing these objectives, and the broader substance and nature of the Fukushima Daiichi nuclear disaster, the investigative field is foremost informed by STS studies. Science, technology and society studies is a multidisciplinary field of research and inquiry that aims to better understand and rethink the relationship among science, technology and society—especially in relation to the political and social, and increasingly, the environmental. STS is thus well placed to address critical problems of the 21st century. Accordingly, STS is a steadily growing area in the social sciences and humanities, globally (CSTMS n.d.). Importantly, with regard to technological development, like nuclear power in Japan, Dutch STS professor Wiebe E. Bijker (2006: 109), outlined:

> Sociological and historical studies have developed a more constructivist analysis of technology in contrast to the standard image of technology that was 'technological determinist'. The resulting social shaping

models stress that technology does not follow its own momentum or a rational goal-directed problem-solving path, but is instead shaped by social factors.

This STS approach (also informed by political science and cultural studies, at the least) well informs the analysis of the Fukushima disaster as a chronic technological disaster, which inadvertently *invited* high vulnerability to natural disaster. In short, STS analysis provides a useful and apt framework for achieving a good understanding of how and why science and technology, here, in the case of nuclear energy development, can be successful or unsuccessful in practice (Bijker 2006: 110).

To arrive at such understanding, to reiterate, the analysis of the book as a whole demonstrates a cartographic approach of mapping and analyzing the sociopolitical terrain of Fukushima as informed by a *policy in action* approach. To engage and inform the reader as to how this terrain mapping and analysis is carried out, at the beginning of each chapter is a brief description of what STS approach has been adopted and what objectives of the book are addressed. Typical questions informing the analysis include: How do we understand the way in which decisions were made concerning nuclear power development, regulation, siting and radiation monitoring—and who represented the driving forces behind the social shaping of these decisions and the attendant developments? What effects did these decisions have on nuclear power development, society and the environment? Who participated in such decision making, and how, and who was left out or marginalized? In the disaster aftermath, what might have assisted governments and industries to manage it better, and what questions are posed regarding future energy pathways and sustainability?

To sum up, in addressing these questions, *Nuclear Disaster at Fukushima Daiichi* presents a powerful, insightful, sometimes provocative, and overall a constructive analysis of critical reflection with coherency and depth. That said, what topics do the contributors address, according to the sequence of the chapters or structure of the book?

THE CONTRIBUTORS' PERSPECTIVES AND TOPICS

The sequence of the chapters reflects the key themes of the disaster as identified by the contributors, and logically begins with local (Japanese) authors mapping out the sociopolitical background that led up to the disaster as a chronic technological one that posed high vulnerability to natural disaster. With the safety of nuclear power plants as the topic of highest regard for both government claims and contestation, the book starts with this theme to critically investigate how nuclear power safety in Japan was socially shaped or projected to suggest or pose a solid foundation for nuclear power development to proceed. The book then builds on this theme by focusing in more detail on the associated safety issue of the siting of nuclear power plants. With safety concerning the actual management and operations of nuclear power breaking down at Fukushima Daiichi, the focus on safety moves postdisaster with the issue of radioactive pollution as an associated facet of disaster, and why and how the Japanese civic sphere contributed to handling it better at the local level and what this might contribute to future disaster management. More critical reflection then occurs on the reasons for and substance of the disaster, and by association what future there is for nuclear power. Finally, the book ends with the proposition that Fukushima will not likely remain the last severe nuclear accident the planet will have to face, and thus a global response for better nuclear disaster management is needed in the form of a global nuclear SWAT (special weapons and tactics) team; of course, not in a law enforcement sense but in the form of an international coordinated emergency response that specifically addresses nuclear disasters. A conclusion then synthesizes the findings and implications and the proffered suggestions of what might best address them.

Nuclear Power Safety and Its Social Shaping

Takuji Hara's contribution in Chapter 2 lies in investigating the social shaping of nuclear safety before and after the disaster—as a key form of STS investigation. Hara explores the question of how and to what extent the belief of safety in nuclear power plants, as shaped by pronuclear interests in Japan, contributed to the Fukushima Daiichi disaster. To determine this he investigates the mechanisms and strategies employed to institutionally and publicly shape this belief through what amounted to quite complex sociotechnological relationships. At the heart of this endeavor was the Genshiryoku Mura, a powerful closed circle of interests directly engaged in nuclear power technology, promotion and business. This revealing account—of the relationships amongst pronuclear government, industry and scientific actors or interests; the material entities of nuclear power plants; and the institutional or structural components of development, regulation and legal systems—clearly makes the case that safety in many ways was compromised to facilitate nuclear power progress, and that here lay a key causal factor of the disaster and the extent of it.

Nuclear Power Siting Problems

In the first chapter of two on this topic, and from the point of view of sociological STS research, in Chapter 3, Kohta Juraku investigates the many structural, institutional and path-dependent background factors informing nuclear power plant siting that appear to have contributed to the Fukushima Daiichi accident and made it worse than it otherwise might have been. A key factor Juraku focuses on was the long-held strategy of siting nuclear power plants in so-called 'nuclear villages'—peripheral or economically disadvantaged rural communities secured through a range of governmental subsidies and incentives in addition to industry ones. Although criticized for many years by citizen activists, journalists, local politicians and academic scholars, the strategy remained virtually unchanged from the beginning of nuclear power utilization, with the promoters of nuclear siting—the national government and utility companies—failing to devise a new siting strategy from the late 1970s. This outdated siting strategy, which led to concentrated siting of multiple reactors at one site as occurred at Fukushima-Daiichi, Juraku argues, led to many negative characteristics of Japanese nuclear power stations in terms of safety that directly contributed to the disaster and the extent of it.

In the second chapter on this topic, Richard Hindmarsh (in Chapter 4) investigates the potent question of the inadequacy of public participation for nuclear power plant siting in Japan—and by implication in other 'nuclear nations'—as a key problem contributing to the Fukushima Daiichi disaster. Hindmarsh locates this question in STS contexts of science, technology and governance, and sociotechnical systems, particularly in relation to major facility siting of megatechnological assemblages that pose high risk of 'radical' change, such as the Fukushima Daiichi disaster put under the spotlight worldwide. This focus is premised by radical change posing significant local controversy for, and highly disruptive impacts on, local place-based communities within which controversial megatechnological assemblages like nuclear power plants are most often sited. In exploring the weaknesses and flawed strategies of Japan's concentrated siting and associated regulatory and public participation policies (which are virtually nonexistent), Hindmarsh, in aligning with the findings of Japanese STS researchers, advances the concept of 'place-change planning'. This concept poses as a facilitative conduit to better and more legitimately engage local place-based communities for more plural, social-situated understandings around where, when and what to site in targeted situations, all of which aims for enhanced social and environmental and energy outcomes.

Civic Radiation Monitoring

On the high-profile postdisaster issue of radiation pollution for citizens, Atsuro Morita, Anders Blok and Shuhei Kimura contribute the first chapter of two in the book (that is, Chapter 5). With a focus on 'emergent concerned groups', these authors investigate the role of civic engagement in the measurement of atmospheric radiation levels, and the stabilization of an emergent measurement infrastructure, as the disaster was actually unfolding. Focusing on the specific case of a 'civic radiation monitoring map' voluntarily organized through ad hoc collaboration among a group of 'amateur' Japanese citizens, they investigate the importance of civic efforts to make radiation data publicly accessible in a situation of widespread public unrest and mistrust in official institutions. As such, Morita, Blok and Kimura advance that this effort served to fill one piece in a serious void of technoscientific credibility left open within Japan following the Fukushima Daiichi disaster. In this sense, the resultant radiation monitoring map, a civic infrastructure, turned out to be an appropriate technological and political response to an emergency situation of extreme uncertainty and lack of reliable information. Accordingly, it highlights how public engagement with technoscience is called for at exactly those moments when established institutional frameworks fail, collapse or otherwise prove inadequate in addressing collectively experienced problems.

In turn, and in demonstrating another civic response to the immediate post-disaster problem of unreliable and often conflicting official radiation data, Denisa Kera, Jan Rod and Radka Peterova (in Chapter 6) investigate the role and impact of the introduction of do-it-yourself (DIY) tools. They relate how DIY Geiger Counters better enabled a grassroots citizen mobilization to measure and share independent radiation data on air, soil, food, and water over the Internet (e.g. on Facebook, YouTube and Google Maps) and mobile platforms—a medium also used to brainstorm what actions citizens might take in their daily lives and neighborhoods concerning this data. In exploring this 'science citizenship', Kera, Rod and Peterova analyze the conditions of and citizen motivations for this mobilization—broadly identified with the so-called Hackerspace movement, open source hardware and the growing use of social networking platforms. In particular, they examine to what extent DIY citizen radiation monitoring played a part in dealing with the complexity and uncertainty of radiation pollution for everyday citizens in the aftermath of Fukushima Daiichi. They advance that such citizen empowerment, which reflects 'cosmopolitical' citizenship, emerged to pose constructive participatory governance lessons for more effective disaster management in the future.

Fukushima as an 'Envirotechnical' Disaster

In Chapter 7, Sara Pritchard's contribution to what occurred at Fukushima Daiichi and all that transpired in the months afterward employs literature and concepts from STS and the history of technology and environmental history. After discussing Perrow's 'normal accidents' and Hughes's 'technological systems', Pritchard conceptualizes and uses an 'envirotechnical systems' approach to analyze the operating procedures at Fukushima Daiichi, as well as emergency measures taken during the crisis. She subsequently argues that environmental factors such as radioactive elements, water, air and human bodies are critical to understanding how the disaster events unfolded. Yet, she warns, there is risk in 'naturalizing' the disaster. Ultimately, Pritchard posits, a complex, dynamic and porous configuration of nature and technology and politics helps us to better understand all that Fukushima now signifies. Her concluding proposition is that 'continued confidence in the ideal of wholly discrete environmental and technological systems is an unfortunate, even dangerous illusion that can only entrench high-modernist hubris and the belief that humanity and technology are separate from the environment'.

The Future of Nuclear Energy

The first of three chapters on the key topic of the future energy implications of nuclear power, as raised by Fukushima Daiichi, is contributed by Catherine Butler, Karen Parkhill and Nicholas Pidgeon. In Chapter 8, these authors point out that nuclear energy has long been a focus of social science research, representing a paradigm case for understanding relations between publics, policy, and science and technologies, including the framing of nuclear safety, and the governance of risk. Following Fukushima, questions about nuclear power were brought to the forefront of debate. In addressing these questions, Butler, Parkhill and Pidgeon examine four broad themes found within existing nuclear energy research: policy, political acceptability and economics; public opinion and attitudes; safety, justice and ethics; and framing and the media. Through interlinked thematic sections of analysis, they provide an insightful overview of central issues and findings in STS and social science research on nuclear energy and then (re)contextualize the key sociopolitical dimensions with reference to the still unfolding disaster in Japan. They conclude that the wider implications of Fukushima—which bring into focus both the fragility of nuclear energy and its durability—in the long term might be found in what arises from the efforts in countries such as Germany to decarbonize without the use of nuclear power.

In the second chapter on this topic, Jim Falk (in Chapter 9) considers the events at Fukushima in the broader Japanese and global system, as both a case study, and a paradigm case, of shifting understandings of 'energy security'. His investigative STS context is informed by the political economy of technological innovation, the dynamics of technological and scientific controversy, the sociology of scientific knowledge, and the social construction of risk and 'risk society'. Nuclear power, Falk argues, has always been enmeshed in a larger set of issues. As such, social constructions of risk and safety, surrounding issues of nuclear weapons proliferation, economics and politics, and the dynamics of technological controversy condition our understanding of the role and prospects of nuclear energy, in addition to contemporary rapid economic, technological, political and ecological transitions. Falk situates Fukushima within this context and posits some observations about the changing relationship of nuclear power to understandings and desires for energy security. A key one is that Fukushima, at the least, will play a strong role not only in future demands of what constitutes safety in nuclear power, but more broadly in the debate about whether, as an institution, the nuclear industry is to be sufficiently trusted.

In the final and third chapter on this topic, Andrew Blowers (in Chapter 10) addresses a fundamental moral question posed by Fukushima: Is it acceptable to continue with nuclear energy? In answering the question, Blowers provides a highly accessible and well informed exploration of the ethical interplay propounded by the arguments for and against continuing with nuclear energy. One such interplay is where those in favor of continuing are those who make the ethical case that the risks from nuclear are insignificant compared to the impending disaster of climate change that nuclear energy might help to avert. Conversely, opponents argue that the risk is not justifiable since there are safer, less costly, more reliable and flexible alternatives available. Overall, Blowers finds that the uncontrolled and unbounded consequences of a nuclear catastrophe like Fukushima suggests nuclear is *morally unacceptable* for the present generation to impose on future generations.

Emergency Response

In a somber closing chapter, Sonja Schmid suggests Fukushima will not likely remain the last severe nuclear accident the planet will have to face. As such, in Chapter 11, she argues that a major shift in our thinking is needed about nuclear risk—away from the failures of accident prevention and toward accident mitigation and more rigorous emergency preparedness. Responding to any future nuclear disasters will thus require much more than retroactive safety adjustments. One proposal Schmid advances, which poses strong traction in the aftermath of the Fukushima disaster, is the creation of a global nuclear emergency response team. But this idea faces a number of challenges, including ones of technical interoperability, national sovereignty, funding and recruitment. Finally, to be most effective, Schmid argues the formation of such a global nuclear SWAT team would need to address fundamental questions of legitimacy, expertise and public trust.

In Conclusion

Following these chapters a summary Conclusion—'Fallout from Fukushima Daiichi: An Endnote'—is provided. It sums up and comments on the findings of, and implications raised by, the contributors and also serves to close the book. That said, to recap, *Nuclear Disaster at Fukushima Daiichi* aims to facilitate wide dialogue to better understand and respond to the Fukushima disaster, for knowledge contribution and social and policy learning, in Japan and globally, for overall, long-term social and environmental well-being and sustainability, particularly in relation to energy choices, safety and futures, and overall to good governance of technology, society and the environment.

REFERENCES

Amos, J. "Fukushima Fish Still Contaminated from Nuclear Accident." *BBC News*, October 25, 2012, accessed October 30, 2012. http://www.bbc.co.uk/news/science-environment-19980614.

Andrée, P. "The Biopolitics of Genetically Modified Organisms in Canada." *Journal of Canadian Studies* 37, no. 3 (2002): 162–191.

Beck, U. *Ecological Enlightenment: Essays on the Politics of the Risk Society*. New Jersey: Humanities Press, 1995.

Bickerstaff, K., I. Lorenzoni, N. Pidgeon, W. Poortinga, and P. Simmons. "Reframing Nuclear Power in the UK Energy Debate: Nuclear Power, Climate Change Mitigation, and Radioactive Waste." *Public Understanding of Science* 17 (2008): 145–169.

Biello, D. "Japan's Nuclear Crisis Renews Debate over Environment, Health and Global Energy Use." *Health Affairs* 30, no. 5 (2011): 811–813.

Bijker, W. "Science and Technology Policies through Policy Dialogue." In *Science and Technology Policy for Development: Dialogues at the Interface,* edited by L. Box and R. Engelhard, 109–126. London, UK: Anthem Press, 2006.

Blowers, A. "Why Dump on Us? Power, Pragmatism and the Periphery in the Siting of New nuclear reactors in the UK." *Journal of Integrative Environmental Sciences* 7, no. 3 (2010): 157–173.

Bradford, P. "The Nuclear Landscape." *Nature* 483, March 8 (2012): 151–152.

Brumfiel, G. "Fallout Foreniscs Hike Radiation Toll." *Nature* 478 (October 27, 2011): 435–436.

Buesseler, K. "Fishing for Answers off Fukushima." *Science* 338 (October 26, 2012): 480–482.

Chan, C-C., and Y. Chen. "A Fukushima-Like Nuclear Crisis in Taiwan or a Non-nuclear Taiwan?" *East Asian Science, Technology and Society: An International Journal* 5 (2011): 403–407.

Cochran, T., and M. McKinzie. *Global Implications of the Fukushima Disaster for Nuclear Power*. Natural Resources Defense Council, 2012.

Commission of the European Communities (CEC). *European Governance: A White Paper*. Com (2001): 428. Brussels: CEC, 2001.

Connor, C. "A Quantitative Literacy View of Natural Disasters and Nuclear Facilities." *Numeracy* 4, no. 2 (2011): Article 2. doi: 10.5038/1936–4660.4.2.2.

Corner, A., D. Venables, A. Spence., W. Poortinga, C. Demski, and N. Pidgeon. "Nuclear Power, Climate Change and Energy Security: Exploring British Public Attitudes." *Energy Policy* 39 (2011): 4823–4833.

CSTMS (Center for Science, Technology, Medicine & Society). "PhD Designated Emphasis in STS." University of California, Berkeley: CSTMS, n.d.

Duffy, R. "Déjà Vu All over Again: Climate Change and the Prospects for a Nuclear Power Renaissance." *Environmental Politics* 20, no. 5 (2011): 668–686.

Emspak, J. "Fukushima Radiation Moving across Pacific Ocean." *Huffington Post,* March 4, 2012, accessed October 30, 2012. http://www.huffingtonpost.com/2012/04/03/fukushima-radiation-pacific-ocean_n_1399843.html.

Flynn, J. "Nuclear Stigma." In *The Social Amplification of Risk,* edited by N. Pidgeon, R. Kasperson and P. Slovic, 326–352. Cambridge, UK: Cambridge University Press, 2003.

Foucault, M. *The Will to Knowledge: The History of Sexuality: Volume One*. London: Penguin, 1990 [French version 1976].

Frackler, M. "Report Finds Japan Underestimated Tsunami Danger." *New York Times,* June 1, 2011, accessed October 30, 2012. http://www.nytimes.com/2011/06/02/world/asia/02japan.html?_r=2&ref=world&.

Freudenburg, W. "Contamination, Corrosion and the Social Order: An Overview." *Current Sociology* 45, no. 3 (1997): 19–39.

Friedman, S.M. "Three Mile Island, Chernobyl, and Fukushima: An Analysis of Traditional and New Media Coverage of Nuclear Accidents and Radiation." *Bulletin of the Atomic Scientists* 67, no. 5 (2012): 55–65.

Funabashi, Y., and K. Kitazawa. "Fukushima in Review: A Complex Disaster, a Disastrous Response." *Bulletin of the Atomic Scientists* 68, no. 2 (2012): 9–21.

GlobeScan. "Opposition to Nuclear Energy Grows: Global Poll, London UK." GlobeScan International (June–September 2011), accessed May 27, 2012. http://www.globescan.com/commentary-and-analysis/press-releases/press-releases-2011/94-press-releases-2011/127-opposition-to-nuclear-energy-grows-global-poll.html.

Gramling, R., and N. Krogman. "Communities, Policy and Chronic Technological Disasters." *Current Sociology* 45, no. 3 (1997): 41–57.

Harmon, J. "Japan's Post-Fukushima Earthquake Health Woes Go beyond Radiation Effects." *Scientific American,* March 2, 2012, 1–4, accessed October 30, 2012. http://www.scientificamerican.com/article.cfm? id=japans-post-fukushima-earthquake-health-woes-beyond-radiation.

Hindess, B. "Power, Interests, and the Outcomes of Struggles." *Sociology* 16, no. 4 (1982): 498–511.

Hindmarsh, R. *Edging Towards BioUtopia: A New Politics of Reordering Life & the Democratic Challenge*. Crawley: University of Western Australia Press, 2008.

Hong, S. "Where Is the Nuclear Nation Going? Hopes and Fears over Nuclear Energy in South Korea after the Fukushima Disaster." *East Asian Science, Technology and Society: an International Journal* 5 (2011): 409–415.

IAEA (International Atomic Energy Agency). "IAEA Update on Fukushima Nuclear Accident." Fukushima Nuclear Accident Update Log, April 12, 2011, 4:45, accessed November 3, 2012. http://www.iaea.org/newscenter/news/tsunamiupdate01.html.

Japan Times. "Utilities Hold onto Plans to Build Reactors." *Japan Times*, October 17, 2012a, accessed November 1, 2012. http://www.japantimes.co.jp/text/nn20121017f2.html.

Japan Times. "Work Resumes at Oma Nuclear Plant." *Japan Times*, October 2, 2012b, accessed November 4, 2012. http://www.japantimes.co.jp/text/nn20121002a4.html.

Jasanoff. S. (ed). *Learning from Disaster: Risk Management after Bhopal*. Philadelphia: University of Philadelphia Press, 1994.

Jorant, C. "The Implications of Fukushima: The European Perspective." *Bulletin of the Atomic Scientists* 67, no. 4 (2011): 14–17.

Kakuchi, S. "JAPAN: Mothers Rise against Nuclear Power." *Inter Press Service*, December 21, 2011, accessed October 29, 2012. http://www.ipsnews.net/2011/12/japan-mothers-rise-against-nuclear-power/.

Latour, B. *Science in Action: How to Follow Scientists and Engineers through Society*. Cambridge, MA: Harvard University Press, 1987.

Maeda, R. "Japan Fukushima Probe Urges New Disaster Prevention Steps, Mindset." *Reuters,* July 23, 2012, accessed November 1, 2012. http://in.reuters.com/article/2012/07/23/us-japan-nuclear-idINBRE86M04320120723.

Matanle, P. "The Great East Japan Earthquake, Tsunami, and Nuclear Meltdown: Towards the (Re)construction of a Safe, Sustainable, and Compassionate Society in Japan's Shrinking Regions." *Local Environment: The International Journal of Justice and Sustainability* 16, no. 9 (2011): 823–847.

McCurry, J. "Fukushima Radiation Fears: Children Near Nuclear Plant to Be Given Monitors." *Guardian,* June 28, 2011, accessed October 30, 2012. http://www.guardian.co.uk/world/2011/jun/28/fukushima-radiation-fears-children-monitors.

McCurry, J. "Anti-nuclear Campaigners Launch Japan's First Green Party." *Guardian,* July 30, 2012, accessed October 29, 2012. http://www.guardian.co.uk/environment/2012/jul/30/japan-green-party-nuclear-power?intcmp=239.

Moe, E. "Vested Interests, Energy Efficiency and Renewables in Japan." *Energy Policy* 40 (2012): 260–273.

Nakamura, A., and M. Kikuchi. "What We Know, and What We Have Not Yet Learned: Triple Disasters and the Fukushima Nuclear Fiasco in Japan." *Public Administration Review,* November/December 2011, 893–899.
Nature. "Lessons of a Triple Disaster." *Nature* 483 (March 8, 2012): 123.
Normile, D. "Fukushima Revives the Low-Dose Debate." *Science* 332 (May 20, 2011): 908–910.
"Nuclear Power Plant Accidents: Listed and Ranked Since 1952." *Guardian,* March 18, 2011, accessed November 4, 2012. http://www.guardian.co.uk/news/datablog/2011/mar/14/nuclear-power-plant-accidents-list-rank.
Onose, S. "Japanese Cabinet Decides Policy on Energy and Environment, without Phasing Out Nuclear Power by 2030s." Department of Information & Communication, Japan Atomic Industrial Forum, accessed November 1, 2012. http://www.jaif.or.jp/english/news_images/pdf/ENGNEWS01_1348468758P.pdf.
Parkhill, K., N. Pidgeon, K. Henwood, P. Simmons, and D. Venables. "From the Familiar to the Extraordinary: Local Residents' Perceptions of Risk When Living with Nuclear Power in the UK." *Transactions of the Institute of British Geographers* NS 35 (2010): 39–58.
Patel, T. "Japan-Style Nuclear Safety Errors Abound, French Regulator Says." *Bloomberg Businessweek*, October 29, 2012, accessed October 30, 2012. http://www.businessweek.com/news/2012-10-29/japan-style-nuclear-safety-errors-abound-french-regulator-says.
Schneider, M., A. Froggat, and S. Thomas. "2010–2011 World Nuclear Industry Status Report." *Bulletin of the Atomic Scientists* 67, no. 4 (2011): 60–77.
Soble, J. "Japan Counts the Cost of Nuclear Withdrawal." *Financial Times,* September 16, 2012, accessed October 30, 2012. http://www.ft.com/intl/cms/s/0/9e9466c6-ffee-11e1-a30e-00144feabdc0.html#axzz2AkiNf7qL.
Stratton, A. "Fukushima Disaster Causes Fallout for Nuclear Industry Worldwide." *Guardian,* March 29, 2011, accessed October 30, 2012. http://www.guardian.co.uk/environment/2011/mar/29/japan-nuclear-clegg-energy-policy.
Suzuki, T. "Deconstructing the Zero-risk Mindset: The Lessons and Future Responsibilities for a Post-Fukushima Nuclear Japan." *Bulletin of the Atomic Scientists* 67, no. 5 (2011): 9–18.Tabuchi, H. "Angry Parents in Japan Confront Government over Radiation Levels." *New York Times,* May 25, 2011, accessed October 30, 2012. http://www.nytimes.com/2011/05/26/world/asia/26japan.html?pagewanted=all.
Tabuchi, H. "Japan Backs off Date to End Nuclear Use." *New York Times,* September 20, 2012, accessed November 3, 2012. http://www.bostonglobe.com/news/world/2012/09/19/japan-refuses-adopt-deadline-for-phasing-out-nuclear-power-plants/N3movcg9iT8gNwsc7stePN/story.html.
Tamman, M., B. Casselman, and P. Mozur. "Scores of Reactors in Quake Zones." *Wall Street Journal,* March 19, 2011, accessed November 20, 2011. http://online.wsj.com/article/SB10001424052748703512404576208872161503008.html.
UBS. "Can Nuclear Power Survive Fukushima." UBS Investment Research, Q-Series®: Global Nuclear Power April 4 (2011).
Unger, S. "Moral Panic versus the Risk Society: The Implications of the Changing Sites of Social Anxiety." *British Journal of Sociology* 52, no. 2 (2001): 271–291.
Westlake, A. "Members of Nuclear Regulation Authority Took Money from Utlities, Manufacturers." *Japan Daily Press,* November 4, 2012. http://japandailypress.com/members-of-nuclear-regulation-authority-took-money-from-utilities-manu facturers-0417796.
Yamaguchi, M. "Japan's Cabinet Backpedals on Totally Phasing Out Nuclear Energy; New Regulatory Panel Starts." *Canadian Press,* September 19, 2012, accessed November 1, 2012. http://essentialforbody.blogspot.com.au/2012/09/japans-cabinet-backpedals-on-totally.html.

2 Social Shaping of Nuclear Safety

Before and after the Disaster

Takuji Hara

This chapter investigates how the belief of safety in nuclear power plants was shaped in Japan before the Fukushima Daiichi disaster (March 11, 2011), and how, postdisaster, the exposed ambiguities and fragility of the safety of nuclear plants changed public beliefs to be instead seen as a strong contributing factor of the disaster. Indeed, it is now widely understood that any belief in the safety of nuclear power plants has collapsed, and that this collapse occurred rapidly following the disaster. For example, an opinion poll conducted by the *Asahi Shimbun* (newspaper) June 11–12, 2011, three months after the disaster, found 74% of respondents preferred the denuclearization of power generation (Asahi Shimbun 2011a). Such figures are in contrast to an opinion survey conducted by the Japanese government on October 15–25, 2009, which indicated that 60% of respondents preferred further expansion of nuclear power generation (Cabinet Office 2009). A strong change in opinion thus appears evident.

At the same time, past opinion polls also indicated doubt over whether society actually or strongly believed in the safety of nuclear power plants before the Fukushima disaster. In the same governmental survey of 2009, 54% of respondents expressed anxiety over safety, which aligned to a longitudinal trend, where, in 2005, 66% of respondents expressed anxiety (Cabinet Office 2006). It thus seems many people long supported nuclear power while also holding anxieties about its safety. This ambiguity in belief about nuclear power safety presents a way to better understand the background of the disaster and why it was not avoided on grounds of safety as a key factor and issue of nuclear power development.

To unveil the process of how such an ambiguous belief in safety was shaped, this chapter conducts a historical analysis based on a perspective called the 'social shaping of technology' (SST), which is found within the field of science, technology and society (STS) studies. According to the SST perspective, any technology is shaped socially through interactions among various human actors, material entities and structural factors. As such, the analytical approach of SST aims to clarify the interactive relationships among such heterogeneous elements (Hara 2007, MacKenzie and Wajcman 1999, Williams and Edge 1996). Accordingly, the following account reveals the complex 'sociotechnological' relationships that finally informed the Fukushima Daiichi disaster. These included first, the 'involved actors' (or stakeholders) of electric power companies, governmental offices, nuclear scientists, local communities, mass media, antinuclear activists, and citizens. Second, the material entities of nuclear fuel, water, vessels, buildings, emergency core cooling systems, and emergency diesel electric generators. Third, the institutional or structural factors of regulatory systems, legal systems, subsidies, research grants, sponsorships, employment opportunities, business opportunities, and economic-geographic structures.

In analyzing the social shaping of the belief of safety in nuclear power plants in Japan, the process of how Japanese society was 'shaped' to such a belief of safety *before* the disaster is first investigated. The behavior of actors in the shaping process is analyzed through attention to their own interactions and those with material entities and institutional and structural factors. An overview of social change related to nuclear safety in Japan *after* the disaster follows. In concluding, I suggest the Fukushima Daiichi disaster gives Japanese society opportunity to change the situation significantly on both the safety of nuclear power plants, including the immediate termination of the most riskiest plants, and in shifts to renewable energy.

THE SOCIAL SHAPING OF 'SAFETY' PRE-FUKUSHIMA DAIICHI DISASTER

In 1954, a budget for research on nuclear energy was approved in Japan for the first time since the end of the Second World War. Electricity generation was considered a decisively 'peaceful' application of nuclear power and was strongly promoted by the Japanese government. The first commercial nuclear power plant was the Tokai Nuclear Power Plant, which began operation in July 1966. Indeed, the mid-1960s to the mid-1990s saw a steady increase in the number of nuclear power plants. However, after 1995, the pace of construction slowed down because of a long economic recession and an accompanying decline in demand for electricity. Otherwise, in terms of safety, frequent accidents and incidents lowered the rate of operation and cost-effectiveness of nuclear power generation (Kikkawa 2004, Yoshioka 2011).

The key government offices promoting the development of nuclear power in the early days were the Science and Technology Agency (STA) and the Ministry of International Trade and Industry (MITI). The former was mainly in charge of the research and development (R&D) of nuclear energy projects, including the fast breeder reactor (FBR) and the nuclear fuel cycle. In turn, MITI was mainly in charge of the promotion of commercial nuclear power plants. Nevertheless, the STA and MITI competed for leadership of Japan's nuclear policy in the 1960s to 1970s. The STA gradually lost power because of technical and social deadlocks in its projects, especially catalyzed by a coolant sodium leak accident in 1995 at the Monju Nuclear Power Plant

(Tsuruga, Fukui Prefecture), which led to a minor radioactive release into the environment (Iino 1999). The STA, then in charge of this plant, attempted to cover up the extent of the accident and resulting damage. This was soon exposed, which resulted in considerable damage to the reputation and power of the STA in terms of safety and responsible management (Yoshioka 2011).

The STA was subsequently dissolved in 2001, which coincided with a government agency restructuring. Its power over nuclear energy was allocated to three agencies: the Ministry of Economy, Trade and Industry (METI, formerly MITI), the Cabinet Office, and the Ministry of Education, Culture, Sports, Science and Technology (MEXT). Of these, the METI became the main government body on nuclear policy, which included placing both the Agency of Natural Resources and Energy (ANRE) and the Nuclear and Industrial Safety Agency (NISA) under its wing. A key thrust of the ANRE was to promote the nuclear industry, while the NISA was in charge of safety regulation as a branch of the ANRE (cf. Cabinet Secretariat of Japan 2012). The ANRE as both promoter and regulator of nuclear power attracted intense criticism about conflict of interest well before the Fukushima Daiichi disaster (e.g. Tateishi 2009). Post-Fukushima, the Japanese government decided to dissolve the NISA to hand over its authority to an independent regulatory body to enhance safety monitoring (Cabinet Secretariat of Japan 2012, World Nuclear 2012). However, the NISA still exists at the time of writing (September 2012), as the succeeding regime was not then ready (cf. Mainichi Shimbun 2012b, Mainichi Daily News 2012).

Returning to the early development of nuclear energy, nine of the 10 electric power companies in Japan, with the exception of the smallest—the Okinawa Electric Power Company, rushed into the construction of nuclear power plants in the 1960s. This was linked to the development of light water nuclear reactors in the USA. Both boiling water reactors (BWRs) developed by General Electric and pressurized water reactors (PWRs) developed by Westinghouse were introduced to Japan. Japanese heavy electric machinery manufacturers partnered with American companies: Hitachi and Toshiba with General Electric; and Mitsubishi with Westinghouse. Japanese electric power companies were also divided into these two partnerships. Tokyo Electric Power (TEPCO), Tohoku Electric Power, Cyubu Electric Power, Hokuriku Electric Power, and the Chugoku Electric Power introduced BWRs from the General Electric-Hitachi/Toshiba alliance. In turn, Kansai Electric Power, Hokkaido Electric Power, Shikoku Electric Power, and Kyusyu Electric Power introduced PWRs from the Westinghouse-Mitsubishi alliance. Most commercial nuclear power plants owned by these electric power companies began operating between 1970 and 1994 (Kikkawa 2004, Yoshioka 2011).

Nuclear power plants in Japan were all located in rural areas and were rather concentrated in certain areas (see also Juraku in this book). Multiple plants were built in the same local area and each plant consisted of plural nuclear reactors. Among the 47 prefectures in Japan, 13 have nuclear plants.

The Fukushima prefecture has two plants and 10 of the 54 commercial nuclear power reactors built thus far. This concentration is related to social reasons rather than technical ones. As the majority of people do not want a nuclear power plant near their place of residence, it has been very difficult for electric power companies to keep finding new places to build nuclear power plants. They have thus tended to build new plants close to, or onto, existing ones, as it was much easier to obtain the agreement of local communities hosting the existing ones, that is, until Fukushima Daiichi occurred.

In general, nuclear power plants found their sites in rural areas short of employment and financially disadvantaged compared to urban areas. If residents in a local government area decided to accept a nuclear power plant, the local community would obtain workplaces, customers, business opportunities, and subsidies. In particular, the Three Power Source Development Laws—the Law for the Adjustment of Areas Adjacent to Power Generating Facilities, the Electric Power Development Promotion Tax Law, and the Special Account Law for Electric Power Promotion—promulgated and enforced in 1974, created the system of large subsidies to local government for siting nuclear power plants in local areas (Kikkawa 2004). With the subsidies, various public facilities were built near nuclear power plants. Local government would also obtain tax subsidies and preferential treatment from central government institutions and donations from electric power companies and their affiliated institutions. In this manner, the argument was that poor local communities could survive and even flourish. Subsequently, several field studies and reports of local residents living close to nuclear power plants have found positive attitudes about the plants (e.g. Kainuma 2011, Kamata 2001).

Further, because government subsidies and preferential treatment were often short-lived and as increased public facilities needed large running costs, local areas with nuclear power plants were more open to more plants. In addition, as the safety issue of nuclear power was not a new one for these areas but an additional one, more local support was found than in areas where they would be new (see also Juraku in this book). In other words, as Kainuma (2011) emphasized, the relationship between local communities embracing nuclear power plants and other parts of Japan is complex and reciprocal. For example, it was also clear that the safety of nuclear power plants had to be emphasized to keep persuading neighborhood residents, and the general public, to accept their construction and benefit for electricity generation.

Nevertheless, over time, events began occurring that saw a deeper questioning emerge of developers and other proponents' claims of safety. This perhaps first began in 1973, when a pivotal dispute over safety occurred. A group of residents in Ikata (Ehime Prefecture) brought a lawsuit against the planning approval for construction of the Ikata Nuclear Power Plant, with the Japanese government the defendant (Abe 1978, NHK 2011, Tanakadate 1978, Yoshioka 2011). Also supporting the plant's construction was the Shikoku Electric Power Company, the Ehime prefectural government, the Ikata local government, and other residents of Ikata. In addition, some nuclear scientists favored construction while others were opposed. Those against the plant believed the company's estimate of a 1/10,000 probability that radioactive substance might be released outside in a possible plant accident was too low. This estimation was calculated based on the assumption that a meltdown would not occur because safety devices, including the emergency core cooling system, would come into play in case of an accident, to sufficiently cool the nuclear fuel. The scientists on the plaintiff (opposing) side insisted the safety devices might not work because of potential technical uncertainties. Those on the defendant side also asserted that the probability (1/1,000,000) of a severe accident with a meltdown was negligible when compared to the benefits nuclear power plant would bring. Notably, with regard to the later breakdown of the emergency cooling system at Fukushima Daiichi, this estimated probability was calculated on the basis of the reliability of the emergency core cooling system (NHK 2011).

The risk of earthquakes in Japan was an associated safety issue. Because the Ikata Nuclear Power Plant was to be built close to the Median Tectonic Line, which included active faults, the plaintiff group raised the danger of accidents caused by earthquakes. The defendant group insisted little evidence existed about active faults being near the plant, and that active faults caused earthquakes. Other scientific and technological issues such as the durability of the reactor and its coolant pipes were also discussed. In any case, these issues did not impact on the court, as the political circumstances of the time supported the promotion of nuclear power plants. In particular, the oil crisis of 1973 had created a strong push for nuclear power plants in influential economic circles in Japan. The district court finally turned down the case on the grounds that approval of the construction of nuclear power plants was informed by the discretion of the Japanese government, as safety issues on nuclear power plants required high-level expertise and were related to broader national policies. The plaintiff made an application for appeal to a higher court, but it was turned down in 1984. A final appeal was made to the Supreme Court, but this was also dismissed, in 1992. Since then and up to this day, Japanese courts have not changed this position, except for a few adjudications in lower courts that later were overturned in higher courts.

Another early pivotal event, and one more public, occurred the following year in 1974, after the controversy began about planning approval for the Ikata Nuclear Power Plant. This more controversial event featured a minor radiation leak from the Japanese nuclear-powered research ship 'Mutsu'. It received wide media coverage and led to more questions about the safety of nuclear power plants. Such incidents were the catalyst, in 1975, for the first national meeting against nuclear power plants. Held in Kyoto at the newly formed Citizens' Nuclear Information Center, established by nuclear scientist Jinsaburo Takagi, thereafter, an influential antinuclear power activist movement emerged (Yoshioka 2011).

The Japanese government responded to these developments by establishing, in 1975, a governmental committee to review the safety regulatory system of nuclear power. Although some committee members advanced a powerful independent regulatory body like the US Nuclear Regulatory Commission, other members opposed this idea fiercely because it posed a slowdown in the construction of nuclear power plants (NHK 2011). Based on the ensuing report of the committee (Ochi 1978), the Japanese government decided to establish the Nuclear Safety Commission of Japan (NSC), under the control of the prime minister. Nevertheless, the NSC suffered in the area of institutional decision-making power. This was because, in being subservient to the Science and Technology Agency and the Ministry of International Trade and Industry, it could only make recommendations to these bodies in the name of the prime minister. Overall, the NSC had little independent authoritative power as a regulatory body.

This position of weak regulatory power was clearly revealed with the NSC's over-hasty safety declaration of nuclear power plants in Japan after the US Three Mile Island accident in 1979 (Yoshioka 2011). What was made clear was that the NSC, after all, was a constituent of the 'Genshi-ryoku Mura': a powerful 'closed circle' of policy makers, electric utility companies, facility manufacturers, and academics who engage in and support nuclear technology in Japan (see also Morita et al. in this book). Three Mile Island had further impacted on the safety controversy of nuclear power plants in clarifying that a meltdown could occur even at a commercial plant with seemingly top-notch safety devices. In 'controversy management' (see Beder 1991), just two days after the Three Mile Island accident, the NSC declared that technical troubles such as those at Three Mile Island were unlikely to cause a severe nuclear accident in Japan.

The overall assertion of Genshi-ryoku Mura authorities was that nuclear power plants in Japan were safer than those in the US (Kagaku Asahi 1979, Yoshioka 2011). This claim was made despite the string of nuclear incidents in Japan by then that contradicted it. Subsequently, the accident at Three Mile Island receded as a distant event for most people in Japan. A similar response by the government, and accordingly by the people, was repeated after the Chernobyl nuclear disaster in 1986. When antinuclear (or nuclear free) activities reignited, Genshi-ryoku Mura authorities repeatedly and rhetorically stated that such accidents could not happen at a nuclear power plant in Japan, as the reactor type at Chernobyl was different from those in Japan and because the Japanese nuclear safety culture was safer (Yoshioka 2011). Accordingly, the policies for nuclear power in Japan did not significantly change, even after the devastating experience of Chernobyl, and the global attention it attracted.

Abetting this situation was the reluctance of the mass media to probe the safety issue. Electric power companies, each virtually a monopoly in their own jurisdiction, were major advertisers that provided the mass media, for example in 2010, with some US$885 million in revenue (Kambayashi 2011, also Morita et al. in this book). It is also important to note that the basically pronuclear government has strong regulatory rights, including those related to media broadcasting. In addition, the government itself was sometimes a major sponsor of the pronuclear campaign in the media (e.g. Amano 2011, Japanese Government 1977), and through other promotional avenues. For example, the ANRE and the STA—as the two powerful institutions in charge of supporting the merits and safety of nuclear power generation—produced a series of largely one-sided and highly promotional materials in support of nuclear power that could only be regarded as rhetoric and soft-sell propaganda (e.g. Sankei Shimbun 2007). The Atomic Energy Commission (AEC) also emphasized the safety of nuclear power plants through lectures or writings by its members (e.g. Kondo 2009, Oba 2010). Further, the NSC proclaimed the safety of nuclear power plants in Japan, as part of its public relations mission on nuclear power safety, which seemed to contradict its regulatory focus on safety. Clearly, the conflicting priorities demonstrated a governmental bias in both development and regulation. Such bias supports the argument of noted STS scholar Sheila Jasanoff, for example, that scientific advisory committees often work as 'active policymakers' rather than conservative scientific advisers (Jasanoff 1990).

Turning to academics, the majority of those in the disciplines of nuclear engineering and radiology stood for the pronuclear power camp, although they were not necessarily active in the public relations of nuclear power generation (Nemoto 1978). Most 'involved' academics worked as members of the AEC, the NSC, and other governmental committees related to supporting nuclear power in Japan. Both Shunsuke Kondo and Haruki Madarame, the incumbent chairs of the AEC and NSC, respectively (at the time of writing), are former professors of the University of Tokyo, the most authoritative and prestigious university in Japan. A large amount of donations and research funds have been awarded to these scholars and their universities by nuclear-related industries and affiliated organizations (Mainichi Shimbun 2012a, Sasaki 2011). In contrast, the few nuclear scientists—including Jinsaburo Takagi, Hiroaki Koide, and Imanaka Tesuji—who joined antinuclear power activities as leading figures, have been treated less favorably in the academic world of nuclear science in Japan; moreover, the mass media rarely disseminated their criticisms of safety before the Fukushima Daiichi disaster.

In looking at the rhetoric of the Genshi-ryoku Mura, various members emphasized the same arguments and claims about the safety of nuclear power plants with different measures and in different arenas, as found in many publications and websites of relevant governmental bodies, electric power companies, and facility manufacturers. A typical assertion of safety by Genshi-ryoku Mura interests is the following:

> If something unusual should happen at a nuclear power plant, various devices for safety would prevent the deterioration of the situation and the development into an accident, based on the fundamental principle

> "Stop, Cool, and Confine". In addition, sufficient measures against natural disasters such as earthquakes, typhoons, storm surges, and tsunamis are adopted. (NSC 2006, as translated by the author)

This assertion advances safety claims at three levels of defense built into the safety system of nuclear power plants. First, abnormal conditions can be prevented through design margins, interlocking and failsafe safety systems, careful maintenance, and quality assurance. Second, abnormal condition propagation can be prevented by the automatic detection and shutdown systems of a reactor. Third, an abnormal condition with an emergency core cooling system can be mitigated through confinement of radioactivity in a reactor's containment system. The safety claims also provided an explanation regarding quintuple protective barriers for containing radioactivity: (1) fuel pellets; (2) fuel rods; (3) reactor pressure vessels; (4) reactor containments; and (5) reactor buildings.

It is therefore not surprising given the powerful pronuclear development interests and their tactics, that despite rising disputes about the safety of nuclear power plants in the 1970s and 1980s, the number of plants increased steadily. In contestation, the antinuclear power camp was simply unable to gain strong public or institutional support. This was partly because internal technical information about the safety of nuclear power plants was not accessible. In addition, the debate on the safety of nuclear power plants was polarized and conflictual—not constructive—which turned off much of the public. The debate was almost always for or against nuclear power plants rather than, for example, on how to make them safer through mutual understandings. It was also difficult for citizens to freely participate in the safety discussion, not only because the debate was highly technical, but also because a number of powerful institutions were keen on establishing nuclear power plants and did not provide the means for sufficient and open public discussion (see also Hindmarsh in this book: chapter 4). In addition, some who were antinuclear were afraid that if they made their position clear they might suffer disadvantage in their careers. Because of these reasons, many withdrew from the available discussions and activities critical of the safety of nuclear power plants.

Meanwhile, nuclear power plant accidents and incidents were becoming more frequent, as were accompanying 'cover-ups' by electric power companies and relevant organizations, which brought about detrimental impacts on the credibility of those organizations when these deeds were revealed (e.g. Genshi-ryoku Siryo Jyouhou-Shitsu 2002, Takagi 2000). As this became more visible, especially from the mid-1990s, more and more people lost confidence in safety management (Asahi Shimbun 1996, Barnard 2000). But in spite of the increasing criticisms about lack of transparency on the operation of nuclear power plants, information disclosure remained limited (Nakamura and Kikuchi 2011, Yoshioka 2011), as was highlighted anew as the Fukushima Daiichi disaster unfolded.

The paternalistic and 'exclusive' attitude of Genshi-ryoku Mura interests thus undermined several opportunities to lessen the possibility of accidents. For example, the risk of nuclear power plant accidents triggered by earthquakes was repeatedly pointed out by several seismologists and commentators and robustly discussed in some media in the late 1990s to 2000s (e.g. Gempatsu Rokyuka Mondai Kenkyukai 2008, Ishibashi 2012). In 1997, well before the Fukushima Daiichi disaster, seismologist Katsuihiko Ishibashi had even coined the term 'genpatsu-shinsai', meaning 'nuclear power plant-earthquake complex disaster'. This forewarned a domino effect of a major earthquake causing a serious nuclear power plant accident near a major population center, resulting in uncontrollable radiation release and earthquake damage severely impeding the residing population's evacuation.

Also reported, in 2003, was a lawsuit raised by a group of activists against the Chubu Electric Power Company for the suspension of Hamaoka Nuclear Power Plant in Shizuoka prefecture. The plant was built very close to the Suruga Trough—the boundary of the Philippine Sea Plate subducted under the Eurasian Plate—the cause of many large earthquakes in Japan. In the courtroom, the resistance status of the nuclear power plant to seismic shocks was a point of contention. The case lasted a long time (nearly four years) because it stepped into complex scientific and engineering issues that were fiercely contested. The risk of 'station blackout', the loss of off-site AC power and any subsequent failure of emergency power on-site was also discussed. As a defense witness, Professor Haruki Madarame, the incumbent chair of the Nuclear Safety Commission of Japan, stated, in 2007, that station blackout was unlikely, and that it was impossible to 'make anything' if one did not overlook 'every small probability' (Hamaoka Gempatsu Tomeyou Saiban no Kai 2007). Although his statement about the risk of station blackout was not incorrect, there was no open discussion between the two sides about how to deal with it, in the event that it might happen, to improve safety. The court rejected the suit in October 2007, finding that the electric power company had devised proper measures to address safety issues.

Another notable 'nuclear incident' occurred on July 16, 2007, when the Chuetsu offshore earthquake occurred in Niigata prefecture and severely shook the Kashiwazaki-Kariwa Nuclear Power Plant (the world's largest plant at the time), located near the epicenter (Cyranoski 2007). Radioactive water spilled out of the spent fuel pool and a small amount of radioactive material was released into the environment. Fortunately, all working reactors automatically stopped and cooled down. The earthquake did not cause a serious accident at the plant. However, the problem was that the shock of the magnitude 6.8 earthquake was stronger than anticipated in the design of the power plant, which was designed to withstand at maximum a magnitude 6.5 earthquake (Cyranoski 2007: 393).

As *Nature* (2007: 387) reported, this raised questions 'about why this facility was built on a seismically active fault line'; especially with this event being 'the third time in as many years that a nuclear power station in Japan has been subjected to an earthquake more powerful than it was designed to withstand'. The questionable location of this plant was actually raised much earlier. During 1975–1977, the advisory council of Japan's Atomic Energy Commission had discussed the geographic validity of the site selected. Although a seismologist member indicated there was a risk of a larger earthquake occurring than stated in the construction plan, as the fault line near the power plant could be longer than recognized, this opinion was not upheld. Thereafter, a tense argument ensued between the seismologist and another seismologist from a government research institution over the evaluation of the faults (Niigata Nippou 2009). Although this point remained unresolved, construction was approved in 1977.

But the matter did not end there. In 1979, a group of worried residents living near the planned construction site brought a lawsuit against the government, arguing that approval to construct the nuclear power plant should be revoked. The 30-year lawsuit included being turned down by the district court in 1994, a follow-up intermediate appeal being turned down in 2005, and a final appeal being dismissed in April 2009, three months before the powerful Chuetsu offshore earthquake shook northwest Niigata. The courts again avoided judgment on the safety of nuclear power plants, and reexamined only the appropriateness of the procedure of governmental safety assessment on the basic design of nuclear power plants.

Two months after this decision, in June 2009, a suspicion about the safety of the Fukushima Daiichi power plant against earthquakes and tsunamis was also raised, in a working group meeting of the Advisory Committee for Natural Resources and Energy by a seismologist from a government research institution. The suspicion was based on the study and record of relevant sediments of the Jogan earthquake and the following tsunami that occurred in AD 869, which was much larger than the earthquake considered in the design of the Fukushima Daiichi power plant (Sougou Shigen Enerugi Chousakai Genshi-ryoku Anzen Hoan Bukai Taishin Kouzou Sekkei Syou-iinkai Jishin Tsunami, Chishitsu Jiban Goudou Working Group 2009). However, the suspicion was not taken seriously by the authorities (NHK 2012). Although the researcher stated that prompt action was needed, any measure to avoid potential damage inflicted by a tsunami was not addressed until it was much too late—that is, after the Fukushima disaster had already occurred.

Clearly, in the history of nuclear power development in Japan, skepticism about, and well-founded research or citizen concerns on, the safety of nuclear power plants was consistently ignored or shunned by the authorities responsible for ensuring adequate safety (see also Takagi 2000).

POST–FUKUSHIMA DAIICHI

After the Fukushima Daiichi disaster, any belief among Japanese citizens in the safety of nuclear power plants collapsed. More than one year after the disaster, 74% of respondents in an opinion poll answered that they did not believe in the safety of nuclear power plants (Asahi Shimbun, 2012). The fragility of a nuclear power plant as a safe and sustainable energy system was thus broadly and decisively recognized. A total station blackout did actually occur and cause reactor meltdown. Everyday people who had been indifferent to or who had ignored nuclear power issues became frightened of radiation exposure by way of air, water, soil, and food contamination. Support for the denuclearization of power generation grew rapidly. The nation's distrust of the government, electric power companies, and relevant academics, who had repeatedly asserted the safety of nuclear power plants in Japan before the disaster, escalated.

Less assertive were the mass media, which adopted an ambivalent attitude. On one hand, media blamed the authorities for an inability to avoid nuclear accidents (e.g. Asahi Shimbun 2011b, Mainichi Shimbun 2011, Yomiuri Shimbun 2011). On the other hand, it seemed they attempted to help the authorities calm people down and divert their attention from the safety of nuclear power plants, arguably by diluting a thorough pursuit of the structural problems of this issue. This lack of probing suggests it as being a residual part of the cozy relationship between some mass media and electric power companies and the government.

The attitudes of local communities close to the Fukushima Daiichi and other nuclear power plants were ambivalent, too. In the months following the disaster, all 50 of Japan's nuclear reactors were shutdown for safety checks (not including the four damaged reactors at Fukushima Daiichi, which are to be abolished) and had to undergo the first-stage test—a design-based simulation test—of a two-phase stress test (cf. World Nuclear News 2012) before they could be restarted. However, without the approval of the local government of any precinct in which a nuclear power plant (or for that matter, any major facility) is located, it is practically impossible to restart the plants even if they were to pass the stress test. The agreement of the local government is influenced by the voice of its constituents including residents. It is uncertain if all temporarily suspended nuclear power plants will be able to obtain the agreement of local governments to restart. Although it is evident that some local communities are economically dependent on nuclear power plants, citizens in these locations are wary after witnessing the damage to local communities that the Fukushima Daiichi disaster inflicted. In addition, the understanding of the geographical scope of a 'local community' has been extended because it is recognized that a broader area than expected can suffer seriously when an accident occurs that involves a nuclear power plant (Daily Yomiuri 2012a), for example, from radioactive pollution.

Nonetheless, electric power companies and powerful economic organizations in Japan requested that the nuclear power plants be restarted, in arguing that power supply and the economy would become unstable otherwise. The government also appears to want to restart the plants (Japan Times 2012b). The campaign to revive nuclear energy in Japan could well be successful, as the resources of the pronuclear power side are much stronger than those of opposing interests. The symbolic event signaling this strength was the restart of Units 3 and 4 of the Oi nuclear power plant in Oi town, Fukui prefecture, as the first restart of nuclear power post-Fukushima. This brokering of strength saw heated arguments over two months involving the Japanese government; the Fukui prefectural government; the Oi town government; other neighboring local governments such as Shiga, Kyoto, and Osaka; Kansai electric power company; national and local business communities; the Nuclear Safety Commission; the Fukui prefectural advisory committee on nuclear safety; antinuclear activist groups; various opinion leaders; and mass media commentators. The Japanese government decided to restart these reactors on June 16, 2012, which occurred the next month. This followed the 'inwardly welcome' agreement of Oi town and Fukui prefecture with the reluctant agreement of neighboring local governments in the Kansai area, where the consumers of electricity generated by the Oi nuclear power plant live (Japan Times 2012c, 2012d, 2012f).

Such intense arguments also reflected the diminishing political power of the Genshi-ryoku Mura, which was on the decline even before the Fukushima disaster. A key factor in this decline was a parallel decline in demand for electric power because of a structural change in Japan's economy; from manufacturing to services and the diffusion of energy-saving technologies in various fields. The cost-effectiveness of nuclear power plants is further questionable when the hidden costs in the back end—that is, the costs needed for dealing with the used nuclear fuel by reprocessing or custody—and the public relations of electric companies and government bodies and subsidies paid to local governments are taken into account (Kikkawa 2004, Yoshioka 2011).

In addition, the frequent troubles and accidents at nuclear power plants, and subsequent cover-ups by relevant organizations (Takagi 2000: 152), have severely injured the credibility of the Genshi-ryoku Mura. This has also been seen in universities where departments of nuclear engineering have been rebadged as quantum engineering and system sciences (the University of Tokyo); sustainable energy and environmental engineering (Osaka University); quantum science and energy engineering (Tohoku University, Nagoya University); or quantum science and engineering (Hokkaido University). Post-Fukushima, TEPCO has also suffered heavily, not only with the reactor meltdowns and severe damage to the Fukushima Daiichi power plant, but also with its failures to ensure safety and better manage the disaster. Extensive restructuring of ownership and management has subsequently occurred through the strong intervention of the national government (Daily Yomiuri 2012b). The NISA and the NSC are also to be restructured into a more independent regulatory body (Mainichi Shimbun 2012b). It is obvious the estimate of risk and costs of nuclear power generation has also substantially increased since the disaster. In addressing these risks and costs a much stronger antinuclear energy movement in Japan has emerged (Japan Times 2012e). This all suggests nuclear power generation in Japan may gradually decline.

CONCLUSION

The social shaping of technology perspective and analytical approach adopted in this chapter focused on the interactions among the heterogeneous human actors, the material entities, and the institutional/structural factors directly informing the Fukushima Daiichi disaster. This has facilitated a better understanding of the sociotechnological phenomenon of the disaster from a safety aspect. And proved highly useful to map out the various individual and organizational actors involved in the process of shaping the safety of nuclear power plants in Japan before the Fukushima disaster. The actors involved in this shaping can be seen as roughly comprising three groups. The first and most influential group was the pronuclear power camp that included the Japanese government, electric power companies, plant suppliers, most nuclear scientists/engineers, and those dependent on resources from them, namely, local governments, local communities, mass media, manufacturers, and service industries. At the core of the first group was the Genshi-ryoku Mura, the closed circle of interests directly engaged in nuclear power technology and business. The second group was the antinuclear power camp comprised of a small number of nuclear scientists and engineers, other antinuclear power activists, a small minority of local residents, and other civic supporters. The third group was the seemingly 'indifferent' general public, but where most became actively NIMBY (not in my backyard) orientated (cf. Tanaka 2004) when it came to the siting of nuclear power plants.

Regarding the influence of the first group, a number of institutions played a key role in justifying the construction of nuclear power plants and in reproducing the supportive social structure for development and expansion. These institutions represented an assemblage of the STA, ANRE, AEC, NSC, and NISA (as organizational actors with governmental authority). They arranged relevant laws and regulations; business alliances between US and Japanese plant suppliers; and government subsidies to coax local communities near nuclear power plants to a supportive position of nuclear power, complemented by donations from electric power companies to, and employment opportunities for, these local communities, as well by plant suppliers and their subcontractors. In addition, there were grants from government and electric power companies to academics; sponsorships and lucrative advertising revenue from electric power companies to the mass media; judicial precedents in favor of progressing nuclear power; and the active use of public relations reflecting rhetoric and soft-sell propaganda to influence social acceptance of nuclear power through a variety of avenues including the media, special journals, forums, and other 'educative' outlets.

Various kinds of material entities were also mobilized and translated to justify the safety of nuclear power plants. Among them, some contributed to the physical isolation of radioactive materials from people, including materials, reactor pressure vessels, containments and buildings, regular cooling systems, water as a coolant, automatic shutdown systems, emergency core cooling systems, other emergency cooling systems, plant system controllers, back-up electric power supplies, emergency diesel electric generators, emergency batteries, safety margins, and so on. Other material entities contributing to the shaping of belief in the safety of nuclear power plants. They included various public amenity facilities and infrastructures built in the local areas hosting the plants, the high walls of nuclear power plants, and TV broadcasting systems.

Also seeking to influence the safety debate, the antinuclear power camp mobilized institutional factors and material entities. Although significantly weaker institutionally and resource-wise than the pronuclear camp, they were arguably more appealing in terms of the high moral ground. This saw organized activist groups; lawsuits, which, although failing in their intent, publicized the issue dramatically; demonstrations and public meetings; antinuclear publications and materials; and, more recently, the Internet employed for information dissemination and activism (see also Morita et al. and Kera et al. in this book).

Occasionally, the mass media did also report oppositional activities as they attracted newspaper sales and TV viewers; and a somewhat balanced engagement was needed in a democratic society to maintain public support and trust. In addition, some staff were sympathetic to the antinuclear power cause. Oppositional activities could also not be ignored by the stronger pronuclear power camp, which was aware that the political power structure might be overturned if the third potentially strong group—the 'indifferent' general public—came to strongly support the opposing camp and thus pressure the government to change its energy policy. Therefore, the pronuclear camp acted strategically to confine and negate the antinuclear power camp. Overall, this attempt was successful before the Fukushima Daiichi disaster but changed dramatically afterwards.

This is because the disaster—stimulated by the natural causes of a huge earthquake and a 'great' tsunami—spoiled and removed the isolating barriers (e.g. the fuel pellets, fuel rods, reactor pressure vessels, reactor containments, and reactor buildings) to both threaten and affect many people with radiation exposure (cf. Pritchard 2012). This acted to significantly dismantle the belief in the safety of nuclear power plants in Japan. Most of the general public came to recognize the danger of nuclear power as a highly risky energy technology. Consequently, Genshi-ryoku Mura appears to have substantially lost authority to promote nuclear power. TEPCO also experienced intense reputational damage, which resulted in extensive restructuring of its ownership and management (cf. Daily Yomiuri 2012b, Japan Times 2012a).

Simultaneously, local communities and governments have increasingly distanced themselves from the pronuclear power camp. The mass media, in keeping with public sentiment and international attention, have increased their criticisms of the government, electric power companies, and mainstream academics in nuclear engineering, although their attitude continues to be seen as ambivalent, as the media also need stable linkages with government and electric power companies as providers of information, licenses, and revenue. At the same time, the presence and strength of antinuclear power activists has increased. More citizens have become sympathetic to the denuclearization movement. Although two of the Oi reactors restarted operation in July 2012, a much stronger antinuclear energy movement is now contesting the field (Japan Times 2012e, 2012g).

Thus, the Fukushima Daiichi disaster, in the long historical context of nuclear power plant mishaps in Japan, finally affected a somewhat dramatic reconfiguration of actors, material entities, institutions, and structures associated with nuclear energy development. However, it is still unclear if such change will endure and for how long. Affecting this equation is the weaker power of the Genshi-ryoku Mura, although most political and economic power holders remain on its side. In this conundrum, denuclearization in Japan will arguably progress slowly. At the same time, the risk of another disaster will not decrease unless public and institutional perceptions of safety change significantly postdisaster.

The perception of nuclear power safety held in Japan before the Fukushima Daiichi disaster was of safety as a 'realistic perfection' with various equipment, rules, procedures, and careful operations. It was projected as being sufficient by the pronuclear camp, which believed that admitting the safety of nuclear power plants was imperfect might hinder the progress of constructing new nuclear power plants or of continuing the operation of existing ones. But, such projections were always impossible because of all the various uncertainties and complexities, including hazardous seismic ones. The pronuclear camp therefore attempted to construct a high belief in safety by mobilizing a package of subsidies, advertisements and public relations, research funds, weak regulation, and political actions. In parallel, opposing views had to be excluded or marginalized as much as possible. This led to the shaping of the Genshi-ryoku Mura community and its attempts to shut down controversy about nuclear energy development (e.g. Beder 1991). However, this social construction of 'safety' could not control the capricious and enormous 'disruptive power' of nature itself, which before the disaster was rather discounted as a background factor in the social shaping of nuclear power.

In conclusion, this study on the shaping process of nuclear power plants in Japan before and after the Fukushima disaster clearly reveals the various pronuclear actors and actions involved in the creation and maintenance of the belief of safety of nuclear power plants. These actors mobilized huge material and institutional resources for this project. However, a significant part of these resources was directed to rhetoric and propaganda (to persuade) or compensation (to avert people's interest) rather than to improve the safety of the technological system itself. In particular, peripheral equipment such as electricity and air supply systems were treated lightly rather than being regarded as critical to safety as a nuclear power plant is a complex and tightly coupled system (Perrow 1984). Turning to the contestational camp, in focusing mainly on the abolition of nuclear power plants, antinuclear activists have neglected the argument for safety improvement, though long and strenuously pointing out the risks of nuclear power plants to the local population and the environment that could be posed by an earthquake, a tsunami, and subsequent station blackout. Such one-sidedness might seem too radical or impractical by some Japanese and undermine arguments for change.

That said, the Fukushima Daiichi disaster now gives Japanese society the required opportunity to change this situation significantly on safety and in shifts to renewable energy (see also Falk, and Blowers, in this book). Postdisaster, more Japanese citizens have become notably interested in the safety of nuclear power plants and the policy of energy supply. To sustain the momentum for change, changing perceptions of safety are key to eliminating the waste of resources currently allocated to nuclear power propaganda and compensation, by which to better improve the safety of nuclear power plants. At the same time, any representation of nuclear risk can inevitably only be limited (Kinsella 2012). Therefore, we can neither say it is safe, nor it must be safe. Instead, we need to adopt a precautionary approach in always regarding nuclear safety as insufficient and requiring continuous care and improvement by applying both rigorous and alternative approaches to safety if we are to retain nuclear power as a future energy source.

REFERENCES

Abe, Y. "Gempatsu Sosho wo Meguru Houritsujo no Mondaiten [Legal Problems about Lawsuits over Nuclear Power Plants]." *Hanrei Times*, 362 (1978): 13–21.

Amano, Y. *Blog*. July 8, 2011, accessed May 21, 2012. http://amano.blog.so-net.ne.jp/2011-07-08.

Asahi Shimbun. "Gempatsu Daijiko "Fuan" ga 73%, Jyouhou Koukai ni Fushin Asahi Shimbun Seron-chousa [Anxious about nuclear accident 73% distrust information disclosure]." *Asahi Shimbun*, March 3, 1996.

Asahi Shimbun. "Gempatsu 'Dankaiteki Haisi' 74% Riyou Sanseiha no 6 Wari mo. Asahi Shimbunsha Seron Chousa [Supporting 'Denuclearization' 74% Including 60% of Supporters of Current Use of Nuclear Power, Opinion Poll]." *Asahi Shimbun*, June 14, 2011a, 1.

Asahi Shimbun. "Shasetsu: Gempatu to Minni, Kimeyou Jibuntachide [Editorial: Nuclear Plants and the Will of the People: Let's Decide by Ourselves]." *Asahi Shimbun*, June 15, 2011b, 12.

Asahi Shimbun. "Oi Genpatsu Saikadou 'Hantai' 54%, Anzensaku 'Shinrai-senu' 78% Asahi Shimbun Seron Chousa ['Opposition' to the Restart the Oi Nuclear Plants 54%, 'Don't Believe' the Safety Measures of KEPCO 78%]." *Asahi Shimbun*, May 21, 2012, 1.

Barnard, C. "The Tokaimura Nuclear Accident in Japanese Newsweek: Translation or Censorship?" *Japanese Studies* 20 (2000): 281–294.

Beder, S. "Controversy and Closure: Sydney's Beaches in Crisis." *Social Studies of Science* 21 (1991): 223–256.

Cabinet Office. "Enerugi ni Kansuru Seron Chousa [The Poll on Energy]." March 13, 2006, accessed May 21, 2012. http://www8.cao.go.jp/survey/h17/h17-energy/index.html.

Cabinet Office. "'Genshi-ryoku ni Kansuru Tokubetsu Seron Chousa' no Gaiyou [Overview of 'The Special Poll on Nuclear Power']." November 26, 2009, accessed May 21, 2012. http://www8.cao.go.jp/survey/tokubetu/h21/h21-genshi.pdf.
Cabinet Secretariat of Japan. "Reform of Japan's Nuclear Safety Regulation." January 3, 2012, accessed May 21, 2012. http://www.cas.go.jp/jp/genpatsujiko/info/kokusaiws/siryo/reform_of_regulation.pdf.
Cyranoski, D. "Quake Shuts World's Largest Nuclear Plant." *Nature* 448, no. 7152 (July 26, 2007): 392–393.
Daily Yomiuri. "Surrounding Govts Want Say in Oi Restarts," *Daily Yomiuri*, April 16, 2012a.
Daily Yomiuri. "Utility Shareholders Meet / Nuclear Power Decried; TEPCO's De facto Nationalization Ok'd." *Daily Yomiuri*, June 28, 2012b.
Gempatsu Rokyuka Mondai Kenkyukai [Research Group of Aging Nuclear Plants]. *Marude Gempatsu nado Naikanoyouni [As If There Were No Nuclear Power Plants]*. Tokyo: Gendai-shokan, 2008.
Genshi-ryoku Shiryo Jouhou-shitsu [Citizen's Nuclear Information Center]. *Kensho Touden Gempatsu Toraburu Kakushi [Investigation on the Cover-ups of Troubles at Nuclear Power Plants by TEPCO]*. Tokyo: Iwanami-shoten, 2002.
Hamaoka Gempatsu Tomeyou Saiban no Kai [The Association for Stopping Hamaoka Nuclear Power Plants]. "(Dai 1 Shin) Dai 8 kai Shonin Jinmon no Boucho Kiroku [An Observer's Record of the Eighth Cross-Examination of Witnesses, the First Trial of the Hamaoka Nuclear Power Plants Case]." Feburary 16, 2007, accessed May 21, 2012. http://www.geocities.jp/ear_tn/.
Hara, T. "Kenkyu Apurochi toshiteno 'Gijyutsu no Syakaiteki Keisei' ['The Social Shaping of Technology' as a Research Approach]." *Nempou Kagaku Gijyutu Syakai* 16 (2007): 37–57.
Iino, H. "Fire by Sodium Coolant Leak at Prototype Fast Breeder Reactor, Monju." Online Ethics Center for Engineering and Research, National Academy of Engineering of the National Academies. 1999, accessed June 29, 2012. http://www.onlineethics.org/cms/22335.aspx.
Ishibashi, K. *Gempatsu Shinsai [Nuclear Disaster Caused by Quake]*. Tokyo: Nanatsumori-Shokan, 2012.
Japanese Government. "Shigen Yugen Jidai no Gensi-ryoku: Isogareru Gempatsu Kensetsu, Anzensei ni Rikai wo [Nuclear Energy in the Age of Limited Resources: Need to Hurry with the Construction of Nuclear Power Plants. Understand Their Safety]." *Economist* (Japan), November 8, 1977.
Japan Times. "TEPCO Asks for Another ¥1 Trillion Injection." *Japan Times,* March 30, 2012a.
Japan Times. "Oi Reacters' Restart Is Vital, Noda Stresses." *Japan Times,* June 9, 2012b.
Japan Times. "Reactors at Oi Plant to Be Reactivated." *Japan Times,* June 17, 2012c.
Japan Times. "Restarted Oi Reactor Generating at Full Capacity." *Japan Times,* July 10, 2012d.
Japan Times. "Massive Tokyo Rally Decries Atomic Power." *Japan Times,* July 17, 2012e.
Japan Times. "Oi's Reactor 4 Achieves Criticality." *Japan Times,* July 20, 2012f.
Japan Times. "Citizens' Groups Propel Rising Wave of Antinuclear Activism." *Japan Times,* August 19, 2012g.
Jasanoff, S. *The Fifth Branch: Science Advisers as Policymakers*. Cambridge and London: Harvard University Press, 1990.
Kagaku Asahi. "Nihon no Gempatsu wa Hontou ni Anzen ka [Are the Japanese Nuclear Power Plants Really Safe?]." *Kagaku Asahi,* June 1979.

Kainuma, H. *'Fukushima' Ron: Genshi-ryoku Mura wa Naze Umareta-no-ka [A Study on Fukushima: Why the Nuclear Power Village Was Born]*. Tokyo: Seido-sha, 2011.

Kamata, S. *Gempatsu Retto wo Iku [Visiting Nuclear Power Plants in the Japanese Archipelago]*. Tokyo: Shuei-sha, 2001.

Kambayashi, H. "Touden Koukoku to Settai ni Baisyusareta Masukomi Gempatsu Houdou no Butaiura! [Behind the Scenes of Nuclear Power Plant-Related News by Mass Media Bought by Advertisement and Entertainment of TEPCO!]." *Bessatsu Takarajima* 1796 (2011): 50–57.

Kikkawa, T. *Dynamism of Development in the Japanese Electric Power Industry.* Nagoya: Nagoya University Press, 2004.

Kinsella, W. "Environments, Risks, and the Limits of Representation: Examples from Nuclear Energy and Some Implications of Fukushima." *Environmental Communication* 6, no. 2 (2012): 251–259.

Kondo, S. "Wagakuni no Genshi-ryoku Kaihatsu Riyou no Genjo to Kadai [Current Situation and Issues of the Development and Use of Nuclear Power in Japan]." Lecture Presented by the Chairman of the Atomic Energy Commission, Genshi-ryoku Anzen Forum, November 7, 2009, accessed May 21, 2012. http://www.aec.go.jp/jicst/NC/about/kettei/091107.pdf.

MacKenzie, D. and J. Wajcman. "Introductory Essay: The Social Shaping of Technology." In *The Social Shaping of Technology,* 2nd edition, edited by D. MacKenzie and J. Wajcman, 3–27. Buckingham: Open University Press, 1999.

Mainichi Daily News. "DPJ Faces Strife over Appointments to New Nuclear Power Watchdog." *Mainichi Daily News*, August 3, 2012.

Mainichi Shimbun. "'Genshi-ryoku Mura' no Heisateki Taishitsu [The Closed Attitude of 'Genshi-ryoku Mura']." *Mainichi Shimbun,* April 21, 2011, 1.Mainichi Shimbun. "Gempatsu Suishin Kenkyu 11 Daigaku ni 104 Okuen, 06–10 nendo, Kuni, Kigyo Teikyo [10.4 Billion Yen Paid by the Government and Companies to 11 Universities for Research in Nuclear Promotion, 2006–10 Financial Years]." *Mainichi Shimbun,* January 22, 2012a, 10.

Mainichi Shimbun. "Bill to Create New Nuclear Regulatory Body Passes Diet." *Mainichi Shimbun Online,* June 21, 2012, accessed July 13, 2012b. http://mainichi.jp/english/english/newsselect/news/20120621p2g00m0dm007000c.html.

Nakamura, A., and M. Kikuchi. "What We Know, and What We Have Not yet Learned: Triple Disaster and the Fukushima Nuclear Fiasco in Japan." *Public Administration Review,* November–December 2011, 893–899.

Nature. "Editorial: Nuclear Test." *Nature* 448, no. 7152 (July 26, 2007): 387.

Nemoto, K. "Shimbun Kiji oyobi Zasshi Rombun ni okeru Genshi-ryoku Hatsuden no Anzensei Ronso no Naiyo Bunseki [A Content Analysis of Discourses about Nuclear Power Generation in the Newspapers and Journals]." *Nihon Genshi-ryoku Gakkai-shi* 20 (1978): 96–102.

NHK. *Genpatsu Jiko heno Doutei: Kouhen [Way to the Nuclear Accident: the Latter Part]*. ETV Special (a TV documentary). September 25, 2011.

NHK. *Umoreta Keikoku [Buried Warning]*. ETV Special (a TV documentary). January 8, 2012.

Niigata Nippou. *Gempatsu to Jishin [Nuclear Power Plants and Earthquakes]*. Tokyo: Kodan-sha, 2009.

Nuclear Safety Commission (NSC). "Genshi-ryoku Anzen no Torikumi ni tuite [On the Safety Measures of Nuclear Power]." NSC, June 2006.

Oba, M. "Genshi-ryoku Korekara no 50 nen—Nihon no Kadai, Nihon no Yakuwari [Nuclear Power in 50 Years Ahead from Now: Japan's Challenges and Roles]," *Energy* 2 (2010): 2–5.

Ochi, K. "Genshi-ryoku Kihon–hou tou Kaisei-houan no Seiritsu to Genshi-ryoku Anzen IInkai no Hossoku [The Passing of the Amended Atomic Energy Basic Law and the Foundation of the Nuclear Safety Commission]." *Nihon Genshi-ryoku Gakkai-shi* 20 (1978): 804–809.

Perrow, C. *Normal Accidents: Living with High Risk Technologies.*New York: Basic Books, 1984.

Pritchard, S. "An Envirotechnical Disaster: Nature, Technology, and Politics at Fukushima." *Environmental History* 17 (April 2012): 219–243.

Sankei Shimbun, "Special article." *Sankei Shimbun* December 28, 2007, accessed May 21, 2012. http://www.win-japan.org/activity/pdf/other03_sairokunishi.pdf.

Sasaki, K. "Goyo Gakusha ga Uketotta Genshi-ryoku Sangyo no Kyodai Kifukin [Huge Donation That Nuclear Industries Paid to Flattery Scholars]." *Bessatsu Takarajima* 1796 (2011): 102–104.

Sougou Shigen Enerugi Chousakai Genshi-ryoku Anzen Hoan Bukai Taishin Kouzou Sekkei Syou-iinkai Jishin Tsunami, Chishitsu Jiban Goudou Working Group [The Joint Working Group of Quake-Resistant/Structural Design and Geological/Seismic Study, The Nuclear Safety Branch of the Combined Research Committee of Natural Resources and Energy]. "Minutes." June 24, July 13, and November 18, 2009, accessed May 21, 2012. http://www.nisa.meti.go.jp/shingikai/107/3/032/gijiroku32.pdf; http://www.nisa.meti.go.jp/shingikai/107/3/033/gijiroku33.pdf; http://www.nisa.meti.go.jp/shingikai/107/3/037/gijiroku37.pdf.

Takagi, J. *Gempatsu Jiko wa Naze Kurikaesunoka [Why Nuclear Accidents Are Not Eradicated].* Tokyo: Iwanami-shoten, 2000.

Tanaka, Y. "Major Psychological Factors Determining Public Acceptance of the Siting of Nuclear Facilities." *Journal of Applied Social Psychology* 34, no. 6 (2004): 1147–1165.

Tanakadate, S. "Ikata Gempatsu Hanketsu no Mondaiten to Gyosei Sosho no Kadai [Problems of the Judicial Decision at the Ikata Nuclear Power Plants Case and Issues of Administrative Litigation]." *Hougaku Seminar* (July 1978): 18–25.

Tateishi, M. "Gempatsu no Taishin Anzensei wo Kangaeru: Kashiwazaki no Kyoukun wo Fumaete [Thinking the Quakeproof Safety of Nuclear Power Plants: Based on the Lessons of the Kashiwazaki]." *Toshi Mondai* 100 (2009): 82–88.

Williams, R., and D. Edge. "The Social Shaping of Technology." *Research Policy* 25 (1996): 865–899.

World Nuclear News. "New Japanese Regulatory Regime as Restarts Approach." *World Nuclear News* March 26, 2012, accessed August 23, 2012. http://www.world-nuclear-news.org/RS-New_Japanese_regulatory_regime_as_restarts_approach-2603127.html.

Yomiuri Shimbun. "Motareau Genshiryoku-mura [Genshi-ryoku Mura Relying on Each Other]." *Yomiuri Shimbun,* June 15, 2011, 1.

Yoshioka, H. *Simpan Genshi-ryoku no Shakai-si: Sono Nihon-teki Tenkai [The Social History of Nuclear Power: Its Development in Japan, New Edition].* Tokyo: Asahi Shimbun Press, 2011.

3 Social Structure and Nuclear Power Siting Problems Revealed

Kohta Juraku

The Fukushima Daiichi nuclear power station accident on March 11, 2011, is one of the most serious accidents in the history of nuclear power utilization, alongside the Three Mile Island and Chernobyl disasters.[1] The cause of this terrible accident is now under long investigation by accident survey committees,[2] journalists, environmental organizations and scholars, amongst many others, such is the global impact of the disaster. These investigations point out many factors that caused, triggered or contributed to the accident, and some are controversial (as the authors in in this book highlight). For example, postdisaster, on December 26, 2011, the Investigation Committee on the Accident at the Fukushima Nuclear Power Stations established by the Government of Japan, released its Interim Report on the Fukushima Daiichi nuclear accident (ICANPS 2011). The report pointed out both technical and institutional problems that should be investigated and resolved as part of avoiding another 'Fukushima'. It stressed a lack of openness to the Japanese public and the international community about the sequence of the accident, the estimation of the amount and distribution of emitted radioactive substances, and the expected impact on public health and the environment.

In this chapter, I report on the findings of my own investigation that the Fukushima Daiichi accident itself was not predetermined just by 'actions of Nature'—the earthquake and the tsunami—and the power station's technological failure to deal with these actions, but also by many prior sociopolitical factors that contributed to this 'technological' failure. An important factor was the nuclear energy policy that shaped the construction, design and siting of nuclear power stations in Japan.[3] In this context, I focus on the concentrated siting of many reactors at a single site, which was an important factor that made the Fukushima Daiichi disaster worse than it otherwise might have been. In critically reflecting upon these factors, my investigation fits into a key objective of this book: to better understand what led to the nuclear meltdown and how might we learn from this for better managing, designing and/or siting nuclear power stations and other nuclear facilities. I begin my investigation by first explaining my method of analysis, which is followed by the analysis and its findings.

METHOD

Pre–Fukushima Daiichi, many Japanese and international scholars studied historically shaped sociopolitical factors of Japan's nuclear development and utilization. Informing post–Fukushima Daiichi investigation, I propose a different picture to these precedents, which builds upon these earlier factors and which also contributes to the analysis of highly structured technological systems per se. As Miwao Matsumoto (1998, 2009) argued, sociotechnical issues like those of nuclear utilization should be interpreted comprehensively and systematically, as being informed by the interactions among political, industrial, academic, civic, and technology sectors. Matsumoto theorized the advantages of this sociological analysis of science, and demonstrated convincing examples of it applied to reproductive medicine and radioactive waste management, among other topics. The sociology of science is a key aspect of science, technology and society (STS) studies, which informs the thrust of this book.

Precedent works on the history of Japan's nuclear program, however, are not as comprehensive or systematic as Matsumoto's proposal. For example, Yoshioka's (2011) social history, considered the standard framework to understand Japan's nuclear program, appears too macroscopic and institutional to best understand the subtle differences in social arrangements involving, for example, 'actors',[4] their networks and political and economic environments, and people's psychologies and historical and cultural backgrounds. This is especially the case when considering sites for nuclear stations as problematic factors that contributed to the technological failure of Fukushima Daiichi. Such arrangements, for example, underpin the 'success' of establishing sites with multiple nuclear power reactors in some local areas and the 'failure' of establishing 'brand-new' siting in other areas. This provides a better explanation of the concentrated siting of nuclear power reactors at the nuclear power station sites of Tokyo Electric Power Company (TEPCO), including Fukushima Daiichi.

In contrast, microscopic and more ethnographic analysis that involved a regional and environmental focus often put too much emphasis on the *detail* of social arrangements at local areas hosting nuclear power stations. For example, Kainuma's (2011) concept of local areas hosting nuclear power stations as 'nuclear villages' provides an interesting explanation for the reason for the concentration of multiple reactors as one of local willingness, not necessarily reluctance (see also Hara, and Hindmarsh, in this book).[5] Indeed, this explanation seems consistent with the phenomena of increased numbers of nuclear power reactors in any one area that already hosted reactors. But these increases were not a result of new power station sites; instead, they were achieved by the addition of reactors or units to existing stations. However, this explanation still lacks careful interpretation of the institutionally shaped sociopolitical environment centering on communities within the siting areas to better understand why such concentration occurred.

In this context, it is important to examine the Japanese government's policy and financial measures for nuclear siting, as concentrated siting was only possible with institutional support, even though some local communities might have been very positive to the idea of having another standalone nuclear reactor in their locale. This makes it clear that the concept of 'nuclear villages' can be expanded to include additional institutional explanations for a better understanding of why the Japanese nuclear power station fleet expanded its number of reactors mainly by reactor addition at existing sites; an understanding currently insufficient according to the existing literature on this topic. This lack of investigation and understanding in Japan is puzzling. For example, the issue of financial dependence on the nuclear industry in siting nuclear power stations is one of the classic issues informing nuclear controversies internationally, but this was seldom mentioned and discussed in Japan until after the Fukushima Daiichi accident.

To address this knowledge gap, I integrate the two influential (institutional and nuclear village) explanations of Yoshioka (2011) and Kainuma (2011), respectively, as discussed previously, as a more comprehensive and systematic investigative approach within a sociology of science analysis. This, as suggested before, is posited to provide a better understanding of why Japan concentrated nuclear power reactors at single sites (or stations, otherwise known as plants), which clearly diverges from the USA, the original designer of Japanese nuclear power stations,[6] and other nuclear countries. In two sections, I undertake a chronological investigation involving significant sociopolitical factors, before concluding. These factors included rapid development and a concentrated siting strategy and the contribution of micro (local) and macro (institutional) structures.

IN THE 'SHADOW' OF RAPID NUCLEAR POWER DEVELOPMENT: CONCENTRATED SITING

From the late 1960s to 1990s, the Japanese nuclear utilization program enjoyed a sustained period of growth and expansion. The number and total capacity of commercial nuclear power reactors increased steadily. As shown in Figure 3.1, about two reactors went online each year. The number ran up to 54 as of 2010, just before the Fukushima Daiichi accident happened, with a total installed capacity of about 50 gigawatt-electric (GWe) (Genshi-ryoku Handbook Hensyu Iinkai 2007). In 2007, this commercial nuclear power generating capacity was ranked third in the world after the USA and France, and number one among nonnuclear-weapon states (Genshi-ryoku Handbook Hensyu Iinkai 2007).[7]

The following diagram (Figure 3.1) illustrates Japan's expansion of its nuclear power utilization program over time. The program has always been claimed as a great success, that is, until the Fukushima Daiichi accident.

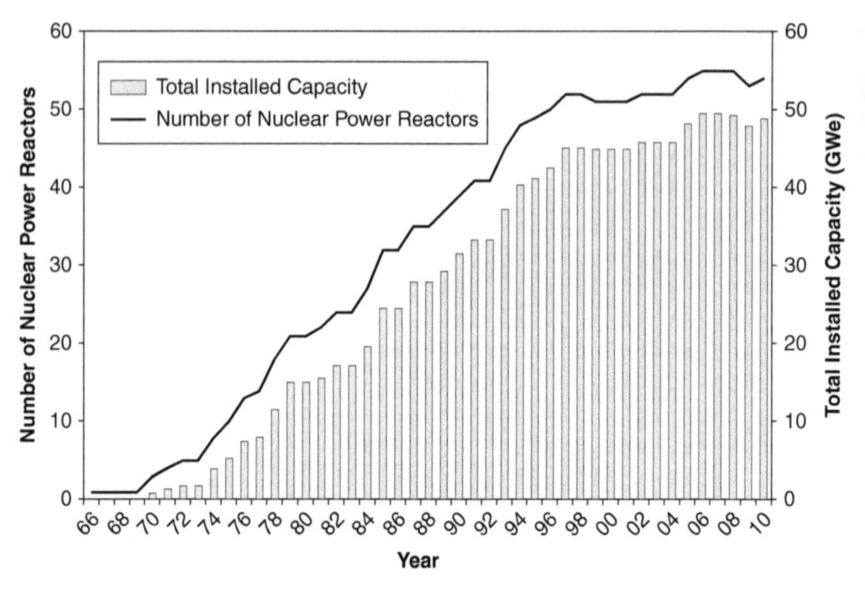

Figure 3.1 Number and total capacity of nuclear power reactors in Japan

Nevertheless, some nuclear stakeholders retain this claim in terms of the program's contribution to the national energy supply, as being convenient and safe, and in contributing to Japan's economic achievements, regardless of the negative impacts of the accident (e.g. FEPC 2012, Miya 2012, Yamana 2012).

Yoshioka (2011) noted the 'great success' of the program was constructed by a highly structured sector of nuclear power interests. Two governmental agencies—the Science and Technology Agency (STA) (which, in 2001, merged with the Ministry of Education, Science, Sports and Culture [MEXT]) and the Ministry of Economy, Trade and Industry (METI, formerly MITI: Ministry of International Trade and Industry)—played key roles in the history of nuclear development (also Hara in this book). But utility power companies, manufacturers, research institutes and universities were also part of an ensuing 'government-industrial complex' (Yoshioka 2011: 24).[8] The Japanese Atomic Energy Commission (JAEC) also played a key role in issuing the 'Long-Term Plan for Research, Development and Utilization of Nuclear Energy' every five years or so. The plan provided the goals that actors (or various players) in the nuclear field had to respond to in developing nuclear power utilization in various fields. These included research and development of advanced nuclear technology, for example, fast breeder reactor development; commercial electricity generation including the expansion of light-water-reactor-based nuclear power generation capacity; and the medical or industrial application of radiation, for example, radiotherapy for the treatment of cancer diseases.

However, Japan's growth in nuclear power capacity demonstrated particular characteristics different from other nuclear power 'aggressive' countries. Figure 3.2 illustrates this difference in adding a third and flatter line to the Figure 3.1 diagram. This flatter line shows the increase in the number of nuclear power station *sites* is much slower than the increase in the actual number of reactors and total capacity, as shown by the steep line. The difference between the two lines well illustrates the unique characteristic of Japanese nuclear siting, as a concentrated siting of nuclear power reactors in fewer power station sites than is the case elsewhere. The number of reactors at any one site internationally is usually up to three or four. In the USA, which has the most nuclear reactors (104) (Denki Shimbun 2010), there is no site that has four or more reactors at a single commercial nuclear power station (RIST 2011).

A key problem of concentration for Japan, though, is very unstable seismic conditions. It is the first major factor I focus on that made the Fukushima Daiichi accident worse than it would have been without such siting concentration. In the accident, the risk of centralized siting of multiple numbers of nuclear power stations and reactors was revealed in a most alarming way, as the disaster unfolded. The Fukushima Daiichi Nuclear Power Station has six reactors located at a single site. Only 12 kilometers south of the Daiichi Station are another four reactors that make up the Fukushima Daini Nuclear Power Station. Both stations are owned and operated by TEPCO. Thus, 10 reactors were located within a 15 kilometer radius and both stations

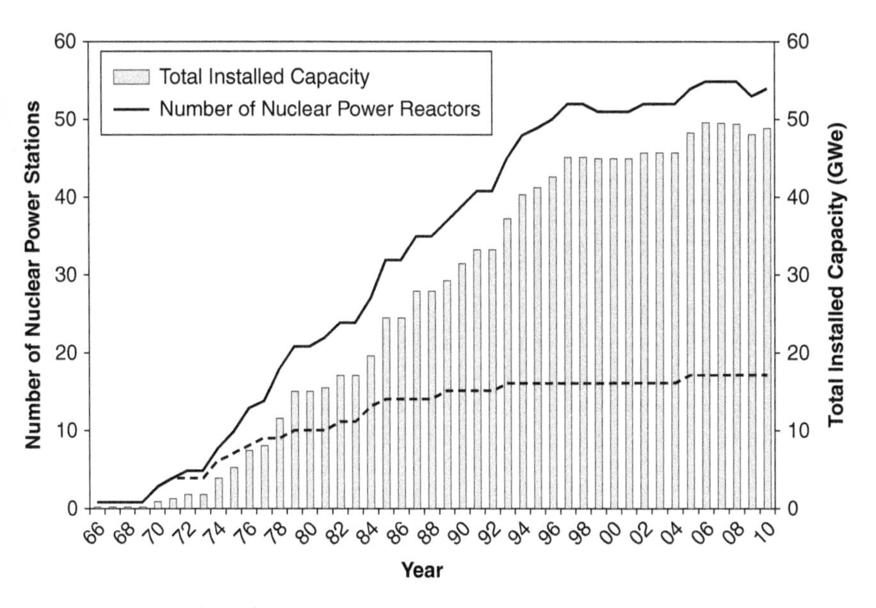

Figure 3.2 Number of nuclear power reactors and stations in Japan

were hit by the same disaster—the magnitude 9 earthquake and subsequent tsunami—which overwhelmed Fukushima Daiichi.

All of the reactors at Fukushima Daiichi were damaged by this natural disaster and had to be taken care of at the same time, that is, securely shut down without too many critical safety components. Each reactor had a spent-fuel pool, which stored used nuclear fuel discharged from the reactor and in some cases 'fresh' fuel for loading. Many of the pools were almost full and needed cooling down continuously to keep their temperatures sufficiently low to ensure safety. The Fukushima Daiichi station is the one of the oldest nuclear power stations in Japan, which meant the station had consumed many spent fuel rods in its operating history. The station thus needed a centralized shared spent-fuel pool, as additional on-site storage, which also had to be cooled down.

In the chaos of Fukushima Daiichi being overwhelmed by the tsunami, TEPCO failed to stabilize three reactors (Unit Nos.1–3), which led to core meltdowns and explosions. Control of the four reactor building spent-fuel pools (for Unit Nos.1–4) was also lost. A number of critical problems thus had to be dealt with simultaneously, and each problem was tied to the others. This domino effect was shown vividly when the almost completed temporarily installed ad hoc external power line to the No. 2 unit was destroyed by the hydrogen explosion of the neighboring No. 1 unit at 3:36 p.m., March 12. This resulted in the delay in activating counter measures to deal with the problems (ICANPS 2011).

Despite this delay, however, the results of the accident were relatively much better than some of the worst-case scenarios that could have occurred if a few things had been different. For example, the counter measures might not have been successful in containing the disaster to these three reactor core meltdowns. If that had occurred, TEPCO most likely could not have avoided a massive fallout of nuclear materials at any of the reactors or spent-fuel pools. All efforts for accident mitigation at all facilities would then have evaporated. No one would have been able to stay on site because of the too-high radiation exposure for human beings. It might even have caused the complete abandonment of all six reactors, or even 10 reactors including the Daini station, depending on the inventory of fallen-out radioactive substances.

According to a newspaper article (Asahi Shimbun 2012), Mr Naoto Kan, the then prime minister of Japan, asked Dr Shunsuke Kondo, then chair of the JAEC and a professor emeritus of the University of Tokyo in nuclear engineering, to estimate the impact of worst case scenarios just after the initial accident happened on March 11 with regard to reactor 1. The simulated response looked catastrophic; a complete evacuation from Tohoku and the greater Tokyo area, of potentially 30 million people, about 30% of the entire Japanese population would have been necessary. Were emergency response crews and the Japanese government ready for this? It is safe to suggest they were not.

That said, it seems clear the centralized co-locating of too many reactors at a single site presented high to extreme risk of vulnerability for so-called common-cause failure events, as the case of Fukushima Daiichi illustrates, although, of course, dependent on geographic location. Centralized co-locating also makes emergency responses extremely difficult to carry out (see also Schmid in this book). Such potential hazards could have been predicted before the Fukushima Daiichi accident in any adequate anticipatory risk assessment of the siting of nuclear power reactors.

In fact, almost 40 years before the Fukushima Daiichi disaster occurred, the famous physicist Seishi Kikuchi encouraged the Japanese nuclear engineering community to implement comprehensive risk assessment of the siting of nuclear power stations, including an upper limit power generation capacity to define an allowable amount of radioactive substances for each reactor and the potential hazard posed of that. The relevant article appeared in 1973 in the *Journal of the Atomic Energy Society of Japan*, the most prestigious and influential academic journal of Japan's nuclear engineering community (Kikuchi 1973). In hindsight, his idea could (and should) have been expanded into a discussion on the upper limit on the number of reactors that could be installed at each site.[9] The second major sociopolitical factor I now address is the contribution of micro and macro structures to this concentration of reactor siting.

MACRO AND MICRO STRUCTURES AS FACTORS UNDERPINNING CONCENTRATION

TEPCO undertook multireactor siting at all of its nuclear power stations. Fukushima Daiichi had six reactors and Fukushima Daini had four, while the Kashiwazaki-Kariwa Nuclear Power Station had seven reactors, making it the world's largest nuclear power station in electricity generation capacity. This station, which is built in close proximity to an active fault line, was itself rocked by an earthquake in 2007, four years before the Fukushima Daiichi incident, thus providing ample time to address the concentrated siting problem. As with Fukushima Daiichi, the quake's magnitude was higher than the Kashiwazaki-Kariwa station was designed to withstand. Although damage to the facility was minor, the incident 'sparked fears that neither Japan's nuclear facilities nor its nuclear safety evaluation system [could] handle the seismic activity that plagues the country' (Cyranoski 2007: 392). Other Japanese nuclear power stations with over four reactors included the Chubu Electric Power Company's Hamaoka station (with five, but two were decommissioned in 2009), and the Kansai Electric Power Company's Oi station (with four). Things changed after the 1970s, though, as Yoshioka (2011:149) outlined:

> There are 16 (nuclear power station) sites in Japan. Almost all of them were announced by power utility companies in the 1960s. . . . There are only two sites of nuclear power stations which were announced

> after 1970s and accomplished until the 2000s: Shikoku Electric Power's Ikata Power Station and Kyushu Electric Power's Sendai Power Station. . . . [M]any siting points . . . were not realized though they were announced by electric companies or negotiated informally. Escalation of the anti-nuclear movement in local areas is the main reason for this difficulty in establishing brand-new sites for nuclear power stations after the 1970s. In particular, many siting projects were deadlocked due to strong opposition by landowners and fishery right holders.[10]

Indeed, all of the currently operating nuclear power stations in Japan found their sites by 1970. The majority received siting permission from 1960 to the early 1980s.[11] After that, things became difficult for any proposed new station siting. For example, in the mid-1980s, in the Maki town case of an attempted nuclear siting, the operator failed to secure the planned land amidst intense local social conflict about safety issues and other points (Juraku et al. 2007). In 2003, siting approval won in 1975 for the Suzu Nuclear Power Station plan—a joint venture of electricity utilities Hokuriku, Chubu and Kansai Electric Powers—was cancelled because of the strong opposition of citizen activists encouraged by the anti-nuclear movement, as particularly triggered by Chernobyl in 1986 (see Yamaaki 2007).

However, this explanation for the collapse of new station siting is insufficient to explain the reason for Japan's concentrated siting, that is, for the co-locating of new reactors at existing station sites. A better explanation for the lack of strong opposition at the concentrated sites is through Kainuma's (2011) 'nuclear village' concept coupled to Yoshioka's explanation. Kainuma illustrates the history of the local communities hosting Fukushima Daiichi and Daini as reflecting the evolution of the nuclear village. These communities embraced nuclear power stations somewhat willingly and spontaneously for the production and reproduction of residents' longing for an improved standard of living. Kainuma found that the majority of local residents never opposed nuclear business and its potential hazards in their daily lives, although this did not mean they were ignorant of the hazards posed. Rather, in being hosts to nuclear power stations, they were proud to achieve a better standard of living, and to support the progress of Japanese society and economy, which had achieved great success in the postwar world.

If we adopt Kainuma's perspective, it can also provide an explanation for the social acceptance of additional reactors at any existing site in difference to those areas which had no existing nuclear facility. That said, what was the institutional explanation for the emergence and evolution of the nuclear village? As Kainuma (2011: 119) explained, financial and employment benefits provided by the nuclear power station itself and the government's subsidy scheme were most influential, even though Kainuma provided the caveat that (local) societies were organized by more than just politics and economy.[12] Yoshioka (2011: 151) elaborates on the subsidy provision as a 'counter-policy against the difficulty of nuclear siting', as an institutional scheme introduced by the national government in the mid-1970s for municipalities to host nuclear stations. Known as the Dengen San-pou Ko-fu-kin Seido scheme (hereafter, the Ko-fu-kin scheme, it was enabled by three acts of parliament on power station subsidies). This scheme then posed an important financial support system to promote economic development for municipalities hosting nuclear power stations. But it also suggested the creation of dependency for these municipalities to continue hosting nuclear power stations and their expansion (Kamata 2001).

The subsidy scheme was established in 1973 as an initiative of then Prime Minister Kakuei Tanaka, an influential conservative politician from Niigata Prefecture. But the original idea for the subsidy scheme came from Jisuke Kobayashi, then Mayor of Kashiwazaki-shi City (Niigata Nippo 2007), the neighboring town to where Tanaka was born. In his term of office, Tanaka promoted the so-called Nippon Kaizo Keikaku (Reformation Plan of Japan). The plan promoted the siting of heavy industry complexes and infrastructure development in rural areas, which included the aim of boosting nuclear power stations for electricity generation. At that time (1973), Japan still suffered a substantial lack of electricity supply, with power source development (installed capacity/planned capacity) at only 44% capacity nationally.[13]

The subsidy scheme also posed as a strategic tool to win local acceptance to host nuclear power stations, as Tanaka was to reveal in a Lower House plenary session of the Diet (parliament): 'Residents of the (siting) area wonder why they have suffer public hazards and other problems due to the electricity supply for the people in big cities, not for themselves. This is a big obstacle to have the consent of local residents' (The House of Representatives 1973).

In addition, the industry minister of his cabinet and one of the initiators of the Japanese nuclear program, Yasuhiro Nakasone, revealed that the subsidy scheme was also needed to boost benefits for local municipalities hosting nuclear power facilities, as they themselves did not provide as much employment opportunity as other industries (The House of Representatives plenary session, May 8, 1973).

In such efforts, Tanaka especially targeted his home area in collaboration with Mayor Kobayashi. A nuclear power siting plan in the Kashiwazaki area was initiated in March 1969 by a city council resolution, although it was put on hold due to tough negotiations with local fishery cooperatives for compensation.[14] Finally, TEPCO, Kobayashi and Tanaka were able to site the Kashiwazaki-Kariwa Nuclear Power Station, with negotiations with the local fishery cooperatives resolved in April 1974, the same year the Ko-fu-kin scheme was established.

Overall, the Kashiwazaki story is an exceptional case of the Ko-fu-kin scheme's effect in establishing the facilitative conduit to add reactors to existing nuclear power stations through the establishment of friendly partnership relationships among municipalities, power utility companies and the national government. Such was its success that the Ko-fu-kin scheme became more generous each year in responding to requests from prefectures and municipalities already hosting nuclear power stations (Shimizu 1991, 1992, Morita 2011).

Such coproduced development clearly demonstrates that nuclear villages coevolved with, and were fostered by, governmental institutional measures. Concentrated siting of nuclear power reactors at each single site of a nuclear power station was a result of this interaction among local areas and the national government and nuclear power developers, through the supportive institutional policy measure of a generous subsidy scheme for already sited areas, in addition to the attempt to foster brand-new siting areas. This created a key component in the 'necessary' background for nuclear power development that indirectly or inadvertently led to the Fukushima Daiichi's superimposed nuclear disaster, which then spoiled the projected history of 'great success' of the expansion of nuclear power generation capacity in Japan.

Supporting the siting strategy were other subtle but influential interactions representing system components that also contributed to the eventual 'grand disaster'. A notable one was the design of legal siting procedures set by the government; another was the regional monopoly system of electric utility businesses, where each area (all 47 prefectures in Japan are divided into 10 areas) was monopolized by a different power company. As Juraku (2011) pointed out, nuclear power station siting processes with a defined legal basis limited the number of stakeholders and the opportunity for them to input their opinions and interests into decision-making processes (see also Hindmarsh in this book: chapter 4). For example, with regard to safety conditions, the legal process requires three de facto conditions of the applicant (i.e. power utility companies) to begin the safety review process, as the only step required to make a legally binding siting permission decision, in the strictest sense. The three conditions are: (1) completion of the planned land acquisition; (2) completion of any required fishery rights compensation agreement—this is because nuclear power stations are all sited on the coastline and their siting obliges fishermen to quit their businesses in surrounding areas, as heated water discharges from nuclear power reactors and/or landfill for reactor sites have a negative impact on local fishing resources (this is why Aldrich [2008] emphasized the importance of winning over fishery cooperatives existing in proposed siting areas); and (3) the consent of prefectural governors and municipal mayors. Basically, this very weak planning approval system for such a potentially risky technology means that applicant power utilities only need to convince landowners, and any fishery cooperatives and elected politicians, in the proposed siting areas of the benefits of the proposed development.

The siting approval process for nuclear power station was, of course, arranged by the national government (i.e. by the Ministry of International Trade and Industry) with the agreement of the power utility industry, which Yoshioka (2011: 161–162) conjectured was to 'promote their [power utilities and MITI's] business in neglecting the escalation of criticisms against nuclear power utilization and of anti-nuclear public opinion'. More officially, it was justified from a development stance informed by the 'Kokusaku Min-ei' (decided by the nation and operated by private enterprise) principle of Japanese nuclear power utilization (Yoshioka 2011: 41). The government defended this arrangement from the key criticism that everything is decided through informal negotiations involving limited stakeholders and a governmental safety review, which never denies license issuance. The government's stance was legitimated (rhetorically) as 'this is a special safety regulation for private business—the electricity power utility business—because of particular potential hazards of nuclear technology. So, we cannot interfere in other parts of their business activities including their negotiation with local stakeholders' (see Juraku 2012: 223).

Notably, through this nuclear power enabling strategy, the 'anti[-nuclear] movement did not have any countermeasures [against the promotion of nuclear siting by MITI and power utilities] because all of the authority to give permission was under their complete control' (Yoshioka 2011: 162). Again, we should remember Kainuma's (2011) concept of a 'nuclear village' in this regard. The limited stakeholder participation in the siting approval process is also consistent with the evolution of nuclear villages. Because it was purposively assumed that the majority of these stakeholders—landowners, local politicians, residents and so forth—had a pronuclear stance, this enabled government and power utilities to promote additional reactor siting without significant obstacles in utilizing this somewhat 'ceremonial' or tokenistic approval process. Mutual collaboration among pronuclear interests in the public and industrial sectors, and social arrangements in local communities, were thus a key factor for concentrated siting to succeed.[15]

CONCLUSION: COEXISTENCE OF 'POLICY SUCCESS' AND SOCIAL FAILURE

Clearly, this story of the structural, institutional and path-dependent background factors—which led to the Fukushima Daiichi disaster being worse than if a more cautious approach had been adopted for nuclear energy development—contradicts or challenges the big success narrative of Japan's nuclear power utilization program. Instead, it reveals and emphasizes narrow-minded plans to achieve policy success for the rapid growth of nuclear power generation capacity, which inadvertently led to downstream policy failure, technically and socially.

The concentrated siting of nuclear power reactors at each single site was seen as a key strategy by powerful pronuclear interests to achieve rapid nuclear utilization. In this strategic agenda-setting context, fundamental problems and issues, which would have hampered progress, were ignored, down-played, neglected or shunted aside. Instead, what was narrowly focused upon by policy makers was the benefit claim of large-scale nuclear power station siting for electricity generation and local economic development (e.g. Irie 1995). Any problems posed were seen as being manageable through both strategic policy development and modification. Overall, the picture emerging is that the nuclear power governance system discounted problems for both strategic and ideological reasons. In turn, this fostered the belief that problematic safety factors would never become critical issues and in any case, they were seen as manageable (see also Hara in this book).

In sum, although acknowledged as potentially possible, these problems were not given critical attention in the policy-making and implementation process. Overlooked was the need for integrated or synergistic planning approaches to best manage uncertainty and complexity to avoid any critical failure of the Japanese nuclear utilization program within a long-term perspective. Instead, a fragmented or reductionist planning approach was adopted that addressed and treated relational system components—including the diverse and important stakeholders affected by concentrated siting—as largely easy-to-manage isolated or separate parts. This was made worse through the actual technical management of the nuclear power station itself including, as in the case of Fukushima Daiichi (see also Pritchard, and Falk, in this book) too many spent-fuels in unprotected multiple-shielded water pools and delays in replacing old reactors (e.g. No. 1 and 2 Units of Fukushima Daiichi). The problem is that all these factors directly and indirectly came together, as they invariably do in sociotechnical systems, to amplify the impact of the Fukushima Daiichi incident. The result was that many publics and economies were adversely impacted by the threat and occurrence of radioactive pollution from the Fukushima Daiichi reactors (Rebuild Japan Initiative Foundation 2012). This presents a massive sociopolitical failure of Japan's nuclear development utilization program, which appears unredeemable.

On a more optimistic note, Japan does have many sociotechnical system concepts, analytical approaches and methodologies incorporated into current decision making. These enable better analysis and understanding of the dynamic and complex interplay found at the intersection of science, technology and society, and the environment. But they are currently underutilized. To avoid another Fukushima disaster, a key policy lesson is that the coexistence of any 'policy success' claims and possible social failure must be considered as an ongoing problematic with regard to advanced technologies, especially highly controversial ones such as nuclear power stations located in highly populated and seismically unstable terrains.

NOTES

Thanks to Jacques Hymans for his comments on an earlier draft of this chapter.

1. Editors note: In this chapter the author (and also Blowers, chapter 10), in comparison to other chapters in the book, refers to nuclear power plants as nuclear power stations. This reflects the use of both terms inside and outside Japan to mean the same thing, where a nuclear power plant or station consists of one or more reactors.
2. The following three committees are recognized as major accident investigative committees: (1) Investigation Committee on the Accident at the Fukushima Nuclear Power Stations, established by the Japanese National Government; (2) Fukushima Nuclear Accident Independent Investigation Commission, established by the National Diet (Parliament) of Japan; and (3) Independent Investigation Commission on the Fukushima Daiichi Nuclear Accident, established by the Rebuild Japan Initiative Foundation.
3. Also see Hara in this book. Matsumoto (2002) theorized the concept 'structural failure' caused by historically shaped social structures at the interfaces of science, technology and society. Matsumoto described this type of failure as one that cannot be explained as natural disaster, nor as one simply caused by human errors, and that we cannot find the root cause of the failure by dichotomist ways of thinking (Matsumoto 2002: 26). Yoshioka (2011) is the most comprehensive social history work on the Japanese nuclear development and utilization problem and provides a number of examples of structural factors embedded in it. Takeda (2011) also illustrated the relationships among political interests, social atmospheres and technological progress. Aldrich (2008) and Juraku et al. (2007) illustrated the effects of structural and institutional factors in the siting of nuclear power stations in Japan. Finally, Kainuma (2011) described the subtle but critical social arrangements that enabled the evolution of nuclear villages, as local siting areas that hosted nuclear power stations without strong negative reactions to nuclear power for many years.
4. Editor's note: 'Actor' (more accurately, 'social actor') is a sociological (or social science) term that can be interpreted as individuals or collectives to the issue or topic under discussion, here, for example, citizens, government agencies, power companies, media and so on, who exercised agency (the social power or capacity to 'act') to actively construct the world, for example, to influence or shape things like policy or people's opinions.
5. It should be noted that the term 'nuclear village' has been used in discussions on Japanese nuclear utilization to criticize the nuclear-industrial complex for many years (Kainuma 2011: 163).
6. Internationally, other nuclear power stations with five or more reactors at a single site include the Bugey and Gravelines nuclear power stations in France and the Bruce and Pickering nuclear power stations in Canada (IAEA 2012).
7. Constructed by the author, based on data found in Denki Shimbun (2010).
8. Yoshioka (2011: 24) defined this concept accordingly: 'In fact, governmental agencies have exclusively made political decisions. Diet (Parliament) has had no actual competence to change decisions made by them or to make their own decision independent from them. Also, the impact of change of government has been relatively low. There have been few cases of political initiatives taken by politicians but influences of them have been diminished by bureaucracy system'.
9. A former vice chairman of the Nuclear Safety Commission of Japan (NSCJ) let the author know about Kikuchi's (1973) paper and expressed his regret about this issue—that is, that open-minded fundamental discussion about

nuclear safety did not follow the publication of Kukuchi's paper in an informal interview conducted by the author in September 2011.

10. Translated into English by the author.
11. Surveyed by the author, based on the data and descriptions in Asahi Shimbun Yamaguchi Branch (2001), Chugoku Electric Power (n.d.), Denki Shimbun (2010), Fukui Atomic Information Center Foundation (2008), Juraku et al. (2007), J-Power (2012), Kyushu Electric Power Company (2012), Matsue-shi City (n.d.), Onagawa-cho Town (n.d.), Shika-machi Town (n.d.), and Tohoku Electric Power Company (n.d.). Japanese siting permission calls for the completion of land acquisition by the nuclear power operator as a requirement. No cases of cancellation have occurred in Japan after siting permission issuance.
12. Translated into English by the author.
13. Statement by Minister Kinji Moriyama at the House of Representatives plenary session, April 11, 1974 (The House of Representatives 1974).
14. Aldrich (2008) illustrates the importance of this compensation negotiation with fishery cooperatives in the whole nuclear siting process.
15. Although I do not discuss it here in any detail, the regional monopoly system and 'fully distributed cost' (FDC) method of cost calculation of electric power utility business in Japan also suggests a strong influence on the steady increase of Japanese nuclear power stations.

REFERENCES

Aldrich, D. *Site Fights: Divisive Facilities and Civil Society in Japan and the West*. New York: Cornell University Press, 2008.

Asahi Shimbun. "Hankei 250 km inai-wo hinan-taisho: Seifu-no 'Saiaku Scenario' [250km Radius Area Might Have to be Evacuated: The Worst Scenario Estimated by the Government]." *Asahi Shimbun*, January 6, 2012.

Asahi Shimbun Yamaguchi Branch. *Kokusaku no Yukue: Kaminoseki Genpatsu Keikaku no 20nen [Trajectory of National Policy: 20 Years of Kaminoseki Nuclear Power Station Plan]*. Kagoshima, Japan: Nanpo-shinsha, 2001.

Chugoku Electric Power Company. "Kaminoseki Genshi-ryoku Hatsudensho: Kensetsu Keikaku: Syuyou Kei-i [Kaminoseki Nuclear Power Station Construction Plan: Outline of Siting Process]." Chugoku Electric Power Company, n.d., accessed February 20, 2012. http://www.energia.co.jp/atom/kami_kensetsu4.html (in Japanese).

Cyranoski, D. "Quake Shuts World's Largest Nuclear Plant." *Nature* 448 (July 26, 2007): 392–393.

Denki Shimbun. *Genshi-ryoku Pocketbook 2010nen-ban [Nuclear Power Pocketbook 2010 edition]*. Tokyo: Nihon Denki Kyokai Shuppan-bu, 2010.

FEPC (Federation of Electric Power Companies of Japan). "Genshi-ryoku Plant Zengouki Teishi-ni-tsuite [About the Shutdown of All of Nuclear Power Plants]." Federation of Electric Power Companies of Japan, May 5, 2012, accessed June 15, 2012. http://www.fepc.or.jp/about_us/pr/oshirase/__icsFiles/afieldfile/2012/05/07/press20120505.pdf.

Fukui Atomic Information Center Foundation. "Fukui-ken no Genshi-ryoku: Bessatsu [Nuclear Power in Fukui Prefecture: Appendix]." Fukui Prefectural Government, 2008, accessed February 20, 2012. http://www.athome.tsuruga.fukui.jp/nuclear/information/fukui/index.html.

Genshi-ryoku Handbook Hensyu Iinkai. *Genshi-ryoku Handbook [Atomic Power Handbook]*. Ohm-sha, 2007.

The House of Representatives (Japan). "Record of the House of Representatives Plenary Session, May 8, 1973." The House of Representatives, 1973.

The House of Representatives (Japan). "Record of the House of Representatives Plenary Session, April 11, 1974." The House of Representatives, 1974.

IAEA (International Atomic Energy Agency). "Power Reactor Information System." International Atomic Energy Agency, 2012, accessed July 19, 2012, http://www.iaea.org/pris.

ICANPS (Investigation Committee on the Accident at the Fukushima Nuclear Power Stations). "Interim Report: Investigation Committee on the Accident at the Fukushima Nuclear Power Stations of Tokyo Electric Power Company," 2011, accessed December 26, 2011 (English executive summary is available at http://icanps.go.jp/eng/interim-report.html).

Irie, K. *Den-en Toshi no Sozo: Dengen Chi-iki Shi-Cho-Son no Arata-na Cho-sen [To Create Power Station Town: New Challenges at Power Station Siting Municipalities]*. Nihon Chiiki Shakai Kenkyu-jo, 1995.

Juraku, K. "Energy-Shisetsu-Ricchi no Shakai-teki Ishi-Kettei Process wo Tou: Koukyou-sei wo meguru Kagaku Gijutsu Shakai-gaku kara-no Approach [Social Decision-Making Process for Energy Facility Siting: An Approach from Sociology of Science and Technology and Public Interests]." PhD dissertation, University of Tokyo, 2011.

Juraku, K. "Genshi-ryoku Hatsuden-sho wo meguru Koukyou-sei to Chiiki-sei [Public Interests, Local Contexts and Nuclear Power Stations]." In Seiyama, K., Ueno, C. and Takegawa, S. (eds.) *Koukyou-Shakai-gaku 1: Risk, Shimin-shakai, Koukyou-sei [Public Sociology 1: Risk, Citizen Society and Public Interests]*, 213–231. Toukyou-daigaku Shuppankai [University of Tokyo Press], 2012.

Juraku, K., T. Suzuki, and O. Sakura. "Social Decision Making Processes in Local Contexts: An STS Case Study on Nuclear Power Plant Siting in Japan." *East Asian Science, Technology and Society: An International Journal* 1, no. 1 (2007): 53–75.

J-Power (Electric Power Development). "Jigyo/Service: Genshi-ryoku Hatsuden Jigyo: Syuyou Kei-i [Business and Service: Nuclear Power Business: Outline of Nuclear Power Business]." J-Power (Electric Power Development), 2012, accessed February 20, 2012. http://www.jpower.co.jp/bs/field/gensiryoku/project/background/index.html.

Kainuma, H. *Fukushima-ron: Genshi-ryoku-mura ha naze umareta-no-ka [Fukushima Studies: Why Did the "Nuclear Village" Emerge?]*. Seido-sha, 2011.

Kamata, S. *Genpatsu-Retto wo Iku [Traveling Nuclear Power Islands]*. Shuei-sha Shinsho, 2001.

Kikuchi, S. "Genshi-ryoku Hatsuden no Anzen-sei to Public Acceptance [Safety of Nuclear Power Generation and Public Acceptance]." *Nihon Genshi-ryoku Gakkai-shi [Journal of the Atomic Energy Society of Japan]* 15, no. 4 (1973): 84–89.

Kyushu Electric Power Company. "Genshi-ryoku Hatsudensho no Shuyou Kei-i [Outline of Nuclear Power Stations Siting Processes]." Kyushu Electric Power Company, 2012, accessed February 20, 2012. http://www.kyuden.co.jp/nuclear_process_index.html.

Mastue-shi City. "Matsue-shi no Genshi-ryoku: Shimane Genshi-ryoku Hatsuden-sho no Gaiyou [Nuclear Power in Matsue-shi City: Outline of Shimane Nuclear Power Station]." Mastue-shi City Hall, n.d., accessed February 20, 2012. http://www.city.matsue.shimane.jp/jumin/bousai/nuclear/gaiyou/gaiyou3–1.html.

Matsumoto, M. *Kagaku Gijutsu Shakai-gaku no Riron [Theory of Sociology of Science and Technology]*. Bokutaku-sha, 1998.

Matsumoto, M. *Chi-no Shippai to Shakai: Kagaku-gijutsu ha naze Shakai-ni-totte-mondai-ka* [*Failure of Knowledge and Society: Why is Science and Technology Problem for Society?*]. Iwanami, 2002.

Matsumoto, M. *Techno-Science Risk to Shakai-gaku: Kagaku-shakai-gaku no Arata-na Tenkai [Techno-Science Risk and Sociology: A New Expansion of Sociology of Science]*. Tokyo: University of Tokyo Press, 2009.

Miya, K. "Han-Genpatsu/Datsu-Genpatsu' Dangi no Mayakashi [Tricks of Anti-Nuclear and Nuclear-Phase-out Arguments]." *ATOMOΣ [Journal of the Atomic Energy Society of Japan]* 54, no. 2 (2012): 1.

Morita, A. "Koueki-zaidan-houjin Genshi-ryoku Kankyo-seibi/Shikin-kanri Center Jutaku-kenkyu Heisei 22-nendo Houkoku-sho [Report for Funded Research by Radioactive Waste Management Funding and Research Center]." 2011, unpublished.

Niigata Nippo. "Yuragu Anzen-shinwa Kashiwazaki-Kariwa Genpatsu: Dai-3-bu Naze Mikai-no Sakyu-chi-ni Dai-4-kai Dengen-sanpou [Questioned Safety Myth: Kasuwazaki-Kariwa Nuclear Power Station: Chapter 3 'Why This Desert Hill Area?' 4 'Three Acts of Parliament on Power Station Subsidies']." December 14, 2007.

Onagawa-cho Town Council. "Genpatsu Shikichi no Kettei kei-i [Process of Site Selection for Nuclear Power Station]." Onagawa-cho Town Hall, n.d., accessed February 20, 2012. http://www.town.onagawa.miyagi.jp/05_04_04_02.html.

Rebuild Japan Initiative Foundation. *Fukushima Genpatsu Jiko Dokuritsu Kensho Iinkai Chousa/Kensho Houkoku-sho [Investigation and Survey Report of the Independent Survey Committee on Fukushima Nuclear Accident]*. Discover 21, 2012.

RIST (Research Organization for Information Science and Technology). "Genshiryoku Hyakka Jiten ATOMICA: Sekai no Genshi-ryoku Hatsuden no Doukou: Hokubei [Nuclear Power Encyclopedia ATOMICA: Nuclear Power Utilization in Foreign Countries: North America]," December, 2011, accessed February 20, 2012. http://www.rist.or.jp/atomica/data/dat_detail.php?Title_No=01–07–05–04.

Shika-machi Town. "Shika-machi no Genshi-ryoku Joho: Syuyou Kei-i [Nuclear Information in Shika-machi Town: Chronology of Shika Nuclear Power Station]." Shika-machi Town Hall, n.d., accessed February 20, 2012. http://www.shika-gen.jp/details.html.

Shimizu, S. "Dengen Kaihatsu Sokushin Taisaku Tokubetsu Kaikei no Tenkai: Genshiryoku Kaihatsu to Zaisei no Tenkai (2) [Evolution of Special Account for Power Station Development Promotion]." *Shogaku Ronshu* 59, no. 6 (1991): 153–170.

Shimizu, S. "Dengen Ricchi Sokushin Zaisei no Chiiki-teki Tenkai [Regional Evolution of Financial Support for Power Station Siting]." *Fukushima Daigaku Chiiki Kenkyu* 3, no. 4 (1992): 3–26.

Takeda, T. *Watashitachi ha kou-shite Genpatsu wo Eranda: Zouho-ban "Kaku" Ron [This Is How We Choose Nuclear Power: Augumented Edition of the "Nuclear" Study]*. Chuko Shinsho La Clef, Chuo Kouron Shin-sha, 2011. (1st edition was published in 2002 as *"Kaku" Ron: Tesuwan-Atom to Genpatsu-jiko no aida ["Nuclear" Study: Between Astro Boy and Nuclear Accident]*) (in Japanese).

Tohoku Electric Power Company. "Higashidori Genshi-ryoku Hatsuden-sho Omona Kei-i [Chronology of Higashidori Nuclear Power Station]." Tohoku Electric Power Company, n.d., accessed February 20, 2012.
http://www.tohoku-epco.co.jp/electr/genshi/gaiyo_higashi/log.html.

Yamaaki, M. *Tamesareta Chiho-jichi: Genpatsu no Dairi-senso ni yureta Noto Hanto Suzu shimin no 13-nen [A Trial against Local Autonomy: 13 years of Suzu Residents in Noto Peninsula with the Proxy War on the Nuclear Siting]*. Katsura-shobo, 2007.

Yamana, H. "Kongo no Kaku-nenryo Saikuru ni-tsuite: Cost tou kara Mita Kaku-nenryo Saikuru [About the Future of Nuclear Fuel Cycle: From the Point of View of Cost Analysis]." *ATOMOΣ [Journal of the Atomic Energy Society of Japan]* 54, no. 4 (2012): 21–27.

Yoshioka, H. *Shin-ban Genshi-ryoku no Shakai-shi: Sono Nihon-teki-tenkai [Social History of Nuclear Power: The Trajectory in Japan]*. 2nd ed. Tokyo: Asahi-shimbun Shuppan, 2011.

4 3/11

Megatechnology, Siting, Place and Participation

Richard Hindmarsh

Nuclear power generation has been in use since the 1950s, and public attitudes have long been subject to change about it. Initial enthusiasm, though ambivalent in many places, has often been replaced by high levels of opposition, as influenced greatly by nuclear power plant accidents, the most damaging being Three Mile Island, Chernobyl and, now, Fukushima (also known as '3/11'). But Fukushima differed from Three Mile Island and Chernobyl as it was situated at the intersection of a natural disaster (here the Tōhoku earthquake and subsequent tsunami)—which the others were not exposed to—and a 'chronic technological disaster'. The latter is informed by human (social) decisions and resulting policies and practices or lack thereof and the discursive interplay of the various stakeholders involved (Gramling and Krogman 1997: 42, see also Chapter 1 in this book for more detail).

Post-Fukushima, the Japanese public recognized the disaster increasingly as one underpinned by social factors, especially sociopolitical ones. Indeed, increasing public distrust of government and regulation to ensure adequate nuclear safety over many decades, alongside the performance of the Tokyo Electric Power Company (TEPCO)—the operator of the Fukushima Daiichi power plant where the accident occurred—to adequately manage the ensuing technological disaster, led to a massive swing of public preferences post-disaster. In contrast to a 2009 Japanese government opinion survey, which found 60% of respondents favored further expansion of nuclear power (Cabinet Office 2009), three months post-Fukushima a leading Japanese newspaper (*Asahi Shimbun*) poll found 74% of respondents favored denuclearization of power generation (see also Juraku, Blowers, Falk, in this book).

This swing also reflected that catastrophic nuclear accidents sit amongst the top of citizen risk perceptions about nuclear activities. They inform widespread feelings of dread of radiation pollution. As such, stress levels rise about potential contamination from living close to a nuclear facility, especially about the potential exposure of children to health impacts (Freudenburg 1997). In turn, fear of place-associated stigmas escalates when nuclear power plants are sited locally in close proximity to a human population, typically a rural community. Stigmatization has a long-lasting effect on residents that goes beyond adverse economic effects such as reduced residential property prices. Local place reputation is at risk, which, following a nuclear facility siting in a particular place, typically sees subsequent avoidance of that place's local products (e.g. food), location or people (Gregory and Satterfield 2002).

Another issue concerns coastal sites chosen for nuclear facilities due to their proximity to electricity grids and seawater for cooling reactors (as at Fukushima Daiichi). As these places are typically already the habitat of human populations and highly valued for recreation, fisheries, iconic seascapes and/or visual impact, the disruption of local ecosystems (e.g. fisheries) through subsequent heated water discharges of coolant seawater can then significantly reduce the attractiveness of these places (Bishop and Vogel 1977: 373). Adverse social, environmental, psychological, cultural and economic impacts can result (e.g. Freudenburg 1997). Overall then, it is not surprising that on the subject of nearby energy, industrial and waste management facilities, citizens rate nuclear facilities the worst (Greenberg et al. 2012).

Residents in affected communities where nuclear facilities are located or planned are therefore most concerned that siting processes 'instil[l] trust and public confidence', and that any facility does not 'impose serious risks to themselves and future generations' (Kunreuther et al. 1990: 481). They are thus keen to have their views included in siting decision making to better check potential for harm and to identify more locally acceptable sites or oppose nuclear power outright for a range of reasons based on local values, knowledges and perspectives. Indeed, such is the dread of nuclear facility siting that Gregory and Satterfield (2002: 355) found that residents want a *high* degree of local participation to overcome any primary control of such decisions by state and federal agencies. After all, they argue it is their 'space', which, endued with their personal meanings has become their 'place' of residence and life experience and perceptions (Vanclay 2008: 4–7). Strong place attachment reflects a broader trend internationally in recent times over disruptive social and environmental impacts of controversial facility siting in local communities, whereby local participatory governance has become increasingly a generic issue for those communities (e.g. Hindmarsh 2010, 2102). In this context, a UK study highlighted a disturbing *lack* of community participation or engagement in relation to nuclear siting decision making (Corner et al. 2011: 4825).

This chapter focuses on this issue of lack of local participatory governance and practice, or, inadequate community engagement by government and developers, for communities faced with nuclear power siting in Japan, more broadly, internationally. As a strategic governmental pathway for nuclear power development, the ensuing analysis finds such inadequate community engagement to be a contributing factor to Fukushima as a chronic technological disaster; that is, in helping to invite (although inadvertently) high vulnerability to natural disaster through flawed top-down siting practices. In its focus, the chapter thus contributes to the book's objectives of contributing to knowledge for policy learning that aligns to long-term sustainability and good governance: to reiterate, as informed by principles of openness, participation, accountability, effectiveness and coherence (Commission of the European Communities 2001) in Japan, regionally, and internationally, here with regard to siting practices of nuclear megatechnologies.

This contribution is, in turn, informed by at least two key themes of science, technology and society (STS) studies—the knowledge field that informs this book (see Chapter 1 for details): those of 'science, technology and governance' and of 'sociotechnical systems'. The first, in the case study of Japan's nuclear power development particularly focuses on good governance for controversial and risky technoscience (e.g. Du Plessis et al. 2010), in addressing 'the ambivalent relationship between democracy and technology' (Nahuis and van Lente 2008: 560). Indeed, the technocratic form of governance—top-down, expert, technophobic and technical—that has characterized nuclear power development and regulation in Japan (e.g. see Hara, Juraka, Morita et al., in this book, also Matanle 2011: 837–838), vividly demonstrates the point of environmental politics scholar Frank Fischer (1990) that technocracy withers democratic government. In this context, Uekoetter's (2012) understanding of the 'authoritarian nature of nuclear technology' is also noteworthy.

The lack of inclusive democratic styles for community engagement about nuclear power plant siting, which Wolsink (2007) also points out with regard to wind energy and many other facility siting examples, can all too easily induce ill-conceived planning outcomes. The latter is posited here with regard to the concentrated siting of six nuclear reactors in one place (see also Juraku in this book), in this case Fukushima Daiichi located on the coastal edge of the Pacific Ring of Fire in the basin of the Pacific Ocean, which has a high density of volcanoes and active faults. Large earthquakes originate where the Pacific Ocean plate is subducted off the east coast of northern Japan (Connor 2011: 2), which is what induced the Fukushima Daiichi tsunami (see Chapter 1 in this book for more detail).

The technical and expert overload in science, technology and environmental policy, planning and management was placed high on the global STS and environmental agenda by sociologist Ulrich Beck. Beck advanced the notion of a 'technical (or ecological) democracy', as informed by his concept of a global 'risk society' (Beck 1992). In a book chapter called 'The World as Laboratory', Beck (1995: 105–110) wrote about the risks of 'technological imperialism' in referring to polluting genetic engineering, nuclear energy and chemical industries and the failures of technical-based regulation. He argued for the need to go beyond technical decision making to a 'political and democratic opening of hazard technocracy' to better prevent technological hazards (Beck 1998: 21). Since then, many new questions about decision-making legitimacy and performance and public trust have been posed in relation to so-called technoscientific risk to the environment, as an inherent part of industrial modernization and its governance. The scope of such risk is indicated by Haraway (1997: 3) who posited, 'technoscience exceeds the distinction between science and technology as well as those between nature and society, subjects and objects, and the natural and artificial that structured the imaginary time called modernity'.

In this context, a key implication highlighted by the Fukushima Daiichi disaster is the increasing need for enhanced public and community engagement for the social appraisal of controversial technoscience, particularly in relation to the special case of megatechnologies like nuclear energy that pose high risk of 'radical change'. This is because megatechnologies feature uncertain, unpredictable and multiple complex interactions unable to be adequately tested in laboratories or computer simulations before their use (Beck 1995: 20). 'Rather, their unanticipated consequences can *only* be discovered *after* they are implemented' (Unger 2001: 282). In turn, 'radical change' is defined here as adverse transformative change which poses significant local controversy to, and impacts on, existing local place-based communities within which controversial megatechnological assemblages most often are sited. This applies to both the social landscape (e.g. to existing social relations and practices of community members and their values) and/or to the surrounding natural landscape. The Fukushima Daiichi nuclear power plant disaster demonstrated all these features for local communities in close proximity to the site of the plant.

Such features also introduce the second STS theme informing this chapter—that of sociotechnical systems as especially characterized by megatechnological assemblages. Sociotechnical systems are seen to represent *tightly coupled* or *co-produced* systems—for example, as applied systems of nuclear power plants and waste repositories, or wind farms, pulp mills, dams, high electricity transmission lines, desalinization plants—which 'link human and social values, behaviour, relationships, and institutions to science and technology . . .' (Miller et al. 2008: 3, also Bijker 1995, Latour 1987, Perrow 1984, Pritchard in this book). Alternatively, sociotechnical systems represent a package of complex social and technical dimensions of megatechnological systems or assemblages, here, the nuclear power plant.

These systems when controversial and posing radical change well conjoin with science, technology and governance issues. This is furthered when their *technical infrastructure* is weakly coupled to or embedded in the broader *social infrastructure*. Weak coupling, in turn, can easily create dysfunctionality and lead to eventual break down (Starr 1999). Many adverse consequences may result, especially when break down occurs chronically. In the case of Fukushima Daiichi, a synergistic natural and technological disaster occurred, the impact of which was then magnified locally for the social infrastructure of communities surrounding the site. For these reasons and more, the analysis of sociotechnical systems is central to the understanding of the nature and dynamics of how to achieve functional systems, such as those pertaining to energy, that pose good governance and sustainability on interrelational social and technical grounds (e.g. Miller et al. 2008). Such coproduction aspects reinforce the need for meaningful, stronger and diverse social input into environmental, science and technology policy and planning. Enhanced public engagement poses as an important conduit for that to occur.

With this focus on the interrelationship between science, technology and governance and (controversial) sociotechnical systems characterized by megatechnology, key questions posed to address the implications of the Fukushima Daiichi disaster include: How did the sociopolitical dynamics of nuclear power development and regulation and inadequate public engagement in Japan contribute to unsustainable energy outcomes (as well indicated in the end by the nuclear disaster)? How was nuclear power development and regulation constructed and maintained to keep public engagement weak? Finally, what change might be suggested to contribute to less risk of disaster and more scope for sustainable energy futures? To answer the first two questions, the ambivalent (or contradictory) relations between nuclear power development (as technology) and public participation (as democracy) in Japan are explored, particularly with regard to the siting of nuclear power plants. The third question is then addressed by suggesting fundamental modification to, or redesign of, existing policy and planning systems for facility siting of (controversial) technological assemblages.

THE AMBIVALENT RELATIONS IN JAPAN BETWEEN NUCLEAR POWER DEVELOPMENT AND PUBLIC PARTICIPATION

During the 1960s and 1970s, the environmental movement emerged globally with not only nuclear waste a key area but also the social responsibility of science movement (led by the UK and USA). The latter movement specifically focused on science and technology and aimed to stimulate public awareness of the social significance of science, and the implications and consequences of technoscientific development according to political, social, environmental and economic factors. Key initial issues were nuclear disarmament and opposition to the bomb. Overall, it was a time when the civic sphere began to actively engage government in demanding more say in how things were being run. In Japan, this coincided with electric power companies developing and operating nuclear power plants finding it increasingly difficult to secure nuclear power plant sites due to growing opposition from local communities—especially landowners and fishery rights holders (Yoshioka 2011: 149)—and citizen activists informed by the steadily strengthening global and Japanese environmental, social responsibility of science, and antinuclear movements (e.g. Avenell 2012, Hara in this book, Ukeoetter 2012). Such opposition in Japan to broader national policies on nuclear energy and electricity production led to a strategic concentrated siting of multiple reactors to bypass the many communities strongly against nuclear power. Siting targeted poor and peripheral local communities where the acceptance of the initial sites for nuclear power was brokered through the introduction of various economic incentives and subsidies, which had induced their dependency as 'one-industry towns' (Nakamura and Kikuchi 2011: 896, also Hara, Juraku, in this book).

In conjunction, a weak regulatory system facilitated the various incentives and subsidies to 'enforce' or best enable later expansion at these sites. Juraku (in this book) outlines that nuclear power plant siting processes limit the number of stakeholders and their opportunity to input their perspectives into decision-making processes, particularly those at the citizen and community level. Only three conditions need to be met to attain siting permission: completion of land acquisition; completion of any required fishery rights compensation agreement (due to the adverse impacts on fisheries from heated water discharges from the seawater cooling systems of nuclear power plants); and the consent of prefectural governors and municipal mayors. To gain support, applicant power utilities have typically targeted local landowners to help win over elected politicians and fishery cooperatives. To facilitate this process, Aldrich (2008: 154–155) found developers

> selected host communities based on the quality and relative capacity of civil society, which aligned to authorities attempting to site reactors in communities with low community solidarity and social capital, and [where] in some cases long-serving legislators would intervene in the siting process to 'bring these facilities into their districts'.

These conditions and processes of siting indicate any meaningful community engagement about siting has either not been encouraged or shut down. More specifically, Fujigaki and Tsukahara (2011: 392) found the official administrative and planning decision-making process for Japanese nuclear power plants to be 'closed off by experts and policy makers, preventing the involvement of resident's in the plants' vicinity'. These authors further remarked: 'In theory, Japan's democratic principles require that the process be open to the public, but in reality . . . there [has been] little effort to include stakeholders in the process of solving complicated problems'. At the same time, developers and government seemingly exploit a 'reluctant acceptance' of nuclear power to justify limited community engagement. For example, a 2004 survey of 1,000 Japanese respondents on public acceptance of nuclear facilities although finding wide 'anxiety' about nuclear power also found a 'necessity of it' for electricity supply, which initially aligned to the reconstruction effort and industrialization of Japan post–World War II (Tanaka 2004)

Weak public involvement in the siting of megatechnologies, furthered by any siting in high-risk terrains, also invites Beck's notion of 'organized irresponsibility'. Japanese reactors are found within the top risk group of very high seismic hazard (Cochran and McKinzie 2012, also Butler 2011). Indeed, of the some 500 operating nuclear power plants worldwide, 24 of the 34 (70%) considered most at risk from hazard because of their proximity to geological faults and the sea are in Japan (Tamman et al. 2011). This number of reactors amounts to roughly 44% of Japan's portfolio of nuclear power plants. Environmentalists and dissenting scientists in Japan have long argued nuclear power plants are just too unsustainable for Japan but their arguments have been ignored (e.g. Avenell 2012: 270). Perhaps incredibly, post-Fukushima, nuclear power proponents in Japan arguing for nuclear power continue to claim it is safe enough (Hara in this book).

Such practices, experiences and developments reinforce arguments for enhanced community engagement and for eliciting valuable local perspectives and social knowledges about where it is best to site or not to site controversial technological facilities, both technically and socially. Benefits of such engagement for nuclear facilities have been found to inform better siting, risk assessment, adverse impact mitigation, public trust and social cohesion, and emergency systems and accident management (e.g. O'Connor and van den Hove 2001). Another benefit, as Beck highlighted in a media interview on the Fukushima disaster (Asahi Shimbun 2011), is that enhanced civic engagement can create spaces for public discussions and the identification of viable and less risky alternatives—including energy efficiency and conservation and renewable energy—as well as further encouraging revitalization of the democratic issue of technology through political mobilization as an important stimulus for policy learning and change (Owens 2004).

Further strengthening the case for enhanced participation are at least *two key problematic themes* found in the literature on concentrated siting of nuclear power plants with relevance to Japan in relation to inadequate public participation: (1) pragmatic reasons for site selection, and (2) new public participation. Again, a key argument is that the scope of the Fukushima Daiichi disaster would have been much less if concentrated siting had not occurred where it did (if at all in Japan given the high risk of seismic hazard), and if enhanced participatory energy governance had been installed in the regulatory apparatus parallel to international trends in progressive public participation practice over the last six decades to inform better siting.

PRAGMATIC REASONS FOR SITE SELECTION

To reiterate, and furthering the detail on the concentrated siting strategy in Japan of nuclear power plant sites in certain locales is their attractiveness to developers because they already reflect 'friendly' ownership of nuclear facilities; have existing infrastructure or potential for easy upgrading; and are situated in communities where public support appears to derive from their familiarity with the industry and financial benefit from it. Typically abetting this support are narratives of safety and benefit, which in the case of Japan is well indicated by Juraku (in this book). A broader international summary of how the nuclear industry rhetorically shapes this benefit image is provided by Blowers (2010: 160):

> [E]mphasis on the positive benefits of nuclear energy almost entirely eclipses the understated risk from nuclear waste which will be present on all these sites. In short, the pragmatic approach to siting focuses on production of energy and the jobs it provides; while the detriments from

long-term highly radioactive wastes on sites, though not denied, are certainly downplayed.

Like the UK, the US also now appears to be actively circumventing potential opposition to new build in 'locating new facilities immediately next to existing ones'. In this context, Fischer's (1999: 296) remark is telling: 'nimby [not-in-my-backyard] has had a profound impact on numerous technological developments, particularly the siting of nuclear power plants and hazardous waste facilities. Indeed, the challenge of nimby is the primary reason for a halt in the siting of new nuclear facilities in the USA over the past two decades'. Over a decade later, Greenberg et al. (2012: 17) would comment: 'Studies have found that those who live near existing energy facilities are more welcoming to new facilities, the so-called "halo" effect which is partly attributable to personal and economic benefits and also to people becoming desensitised to risk'.

The UK Department of Energy and Climate Change in assessing potential sites for additional plants relayed a similar narrative of acceptance or perhaps resignation amongst those targeted: 'People living and working nearby have had a long time to get used to there being an adjacent nuclear plant so this [additional plant] is unlikely to be a problem at this location' (Blowers 2010: 160). Blowers (2010), however, criticized this argument as speculation as no research into local community attitudes, values and opinions on the case for 'new build' had been undertaken.

Juraku (in this book) describes local areas in Japan hosting nuclear power stations as 'nuclear villages' to explain relative local willingness to nuclear expansion—at least before Fukushima Daiichi, with deep skepticism setting in postdisaster across Japan about nuclear power. Such willingness, again, was most evident in peripheral Japanese communities seeking improved standards of living as posed by institutional and developer incentive strategies (Yoshioka 2011, also Morita et al., and Hara, in this book). More implicitly, subsidy programs to promote economic development for municipalities hosting nuclear power stations by the Japanese government were developed as a 'counter-policy against the difficulty of nuclear siting' from the mid-1970s (Yoshioka 2011: 151 cited in Juraku in this book, see also Kikkawa 2004). Obviously such cooptation or manipulation would also serve to counter resistance, in addition to any demands for stronger public engagement.

Nonetheless, in these communities deep opposition was also found; in some instances by citizens who challenged any siting of nuclear power plants in court cases. However, these challenges invariably failed for institutional reasons; typically the courts deferred the matter back to the regulatory authorities (see Hara in this book). But to reiterate, weak public engagement mechanisms largely discouraged contestational mobilization in the first place. Such contestation reflects the broader literature that shows both positive and negative opinions often exist (sometimes simultaneously) for those living in close proximity to nuclear facilities (like other controversial facilities). This ambivalence demonstrates that local people can quite often experience complex and sometimes ambiguous perspectives about the siting of controversial technological assemblages in 'place' (e.g. Pidgeon et al. 2008), which then also informs conditional acceptability. Conditional aspects of public acceptability regarding nuclear power include doubts about safe disposal and storage of waste; concerns about decommissioning; a perceived lack of safety overall; and preferences for other energy options (Corner et al. 2011: 4826). This then reinforces strategic siting assessment processes put forward by government and developers for multisiting at existing nuclear sites to avoid brokering aspects of conditional acceptability.

As such, in targeting peripheral communities on the edge of the mainstream for multisiting, Blowers (2010: 162) summarized the following characteristics for doing so in the UK, which tends to align to the Japanese case and can be seen to further disempower local communities to actively contest, resist or be critical of nuclear facility siting or to seek robust public engagement. First, targeted communities are usually located in areas geographically remote or relatively inaccessible; second, they display economic marginality in places that are monocultural (which are easier to control in contrast to places with more cultural diversity) and dependent on a dominant employer or employment sector; third, they display political powerlessness, which makes it easier to make key decisions elsewhere, such as in centralized government planning or industry spaces often found in distant metro centers; fourth, cultural defensiveness where ambivalent attitudes combine with feelings of isolation and a fatalistic acceptance of nuclear activities; and fifth, environmental degradation in proximity to areas of radioactive contamination or places where high radioactive risk is already present.

However, an associated problem for safety is posed when 'industries may choose to site their facility on the basis of the lack of resistance rather than on the safest environment . . .' (Gramling and Krogman 1997: 4). This would become only too evident with the siting location of the Fukushima Daiichi nuclear power plant and its six reactors.

NEW PUBLIC PARTICIPATION

Historically, as well indicated previously, public participation in nuclear policy making has been severely limited in Japan. In December 1995, pressure for change was catalyzed by the 'Monju' fast breeder reactor accident (Tsuruga, Fukui Prefecture, SW Japan), which, although nonfatal, involved 700 kilograms of sodium leaking from the secondary coolant system to cause a fire. The lack of appropriate response and crisis management by the operator—the Power Reactor and Nuclear Development Corporation—was strongly criticized by the public. After this and several prior nuclear power safety incidents, mounting public distrust of the government over safety issues became too visible to ignore.

To 'manage' these issues and build trust, a roundtable conference (a new participatory tool found in contemporary international trends towards participatory governances) on nuclear policy was utilized by the Japanese government to discuss 'so-called STS issues in Japan' (Juraku et al. 2007). This initiative was 'regarded as an epoch-making event. For the first time in almost 40 years of Japanese nuclear power program history, the government invited several representatives from the anti-nuclear side' (Juraku et al. 2007: 55). However, the conference failed on many counts—a prime one being a lack of direction set by the organizers; further, in an attempt at coproducing an agenda for the roundtable's process, participants could not agree on its goal. Subsequently, the outcomes failed to produce clear decisions on Japanese nuclear policy, and in 1999 the conference was suspended, Since then, the 'failed' initiative 'is often cited as an example of the difficulties of the participatory approach to nuclear issues' in Japan (Juraku et al. 2007: 55), thus also providing justification for the status quo of 'technocratic criteria', that is, 'nonpolitical' technical criteria, to continue dictating site selection (Aldrich 2008: 147). This retention of technical criteria as the most suitable criteria for site selection, however, contradicts a burgeoning literature that finds social criteria are often just as important to include in such decision making as technical criteria, sometimes even more so (e.g. Jones and Eiser 2010).

Signaling the importance of social criteria at the community level in Japan, Juraku et al. (2007) cited strong local citizen initiatives emergent on nuclear power. A key citizen initiative saw the cancellation of a nuclear power plant by a local citizen referendum in Maki-machi in Niigata prefecture; and a plan in the Hokkaido prefecture to build a new plant was approved only after the deliberation of an inquiry committee of citizens and experts. But overall these exercises have been a rarity, primarily due to institutional resistance from centralized nuclear power regulatory and development decision makers (Fujigaki and Tsukahara 2011).

This begs the question: how might local communities in Japan and elsewhere faced with nuclear power siting proposals become better involved in decision making about such facility siting, drawing on the lessons for enhanced community engagement both in Japan and overseas? Given that our focus is on community engagement and nuclear power in reflection of Fukushima it is reasonable to start with Japan. Indeed, in the immediate aftermath of the disaster, Fujigaki and Tsukahara (2011: 392) proposed the need to 'construct a reliable and transparent public sphere in which to discuss and examine . . . future energy policy'.

Along those lines, in relation to local siting decision making this proposal reflected the referral of Juraku et al. (2007: 58) to the Maki-machi case as an example 'to think about the critical conditions of participatory methods in local contexts'. These authors were impressed by the (spontaneous) local and independent self-managed referendum being suggested not by outside experts but by a citizen's group of local residents with no legal authority. Their deliberations influenced the cancellation of the nuclear power siting proposal, which these authors posited amounted to an informal mechanism that might have a 'big influence on the entire decision-making process'. As such, Juraku et al. (2007) advanced that the resulting 'formal-informal social decision-decision making process' in the Maki-machi case provided a flexible arrangement to best address the local political context.

More broadly, the brokering of this arrangement reflects what is known as a 'politics of place', especially 'politics that transgress place' (Urbanik 2007: 1206). Fundamentally, place politics concerns the exercise of territorial power by insider groups located in a given place (e.g. residents), and outsider groups located outside it (e.g. developers, government and external public communities). Insider place politics reflected the citizen referendum initiative that Juraku et al. (2007) cited as offering substance for the 'improvement or revision of the implemented formal processes', to build in flexibility through informal mechanisms 'to adapt to the context and situation of each case' (or local place of siting) (Juraku et al. 2007: 70).

In agreeing with Juraku et al. that the insertion into existing planning systems of a citizen referendum at the local level would engender meaningful dialogue and interaction of local stakeholders and communities for valuable input into facility siting processes, a range of deliberative mechanisms exist to select from for a case-by-case approach. Other mechanisms include inquiry groups, think tank working groups, citizen review panels or open space working groups (e.g. Edwards et al. 2008). Such suggestion also works in well with the more sophisticated concept of 'place-change planning' (following Hindmarsh 2012).

PLACE-CHANGE PLANNING

Place-change planning challenges top-down imposed or suggested radical change posed by siting megatechnologies at the local place-based community level. It does this, as also suggested previously, by creating and inserting deliberative participatory mechanisms into existing policy and planning systems to engender meaningful dialogic interaction of local stakeholders and communities about such change for policy input and influence, according to principles of good governance by which to enhance technological, environmental and social decision making. Pragmatically, such mechanisms aim to best release local social perspectives and knowledge for valuable input on facility siting in contexts of good governance. To do so they aim to create appropriate civic spaces to probe the underlying 'rationality contexts'—that is, the value, belief and attitudinal contexts—of local place-based communities subject to facility siting of a radical nature to better identify and address any potential planning problems/issues posed (Hindmarsh 2010, 2012).

Such rationality contexts reflect, by and large, socially derived knowledge informed by the ways individuals interact with their social environment, either transmitted to an individual by other persons, or constructed by an individual specifically about social phenomena (Turiel 1983: 1–2). For example, socially derived knowledge are informed by the intersection of historical, sociocultural, biophysical and political-economic aspects of any local community (Spink et al. 2010, Woolley 2010). The explicit social environment referred to here is that of the place-based community, and its rationality contexts are those of the individual *within* community (Colclough and Sitaraman 2005). In response to such stimuli and context, place-based perceptions of needs and priorities form and typically become strongly articulated in relation to proposed change in reflecting positions to such change situated along a 'resistance-acceptance spectrum' (Hindmarsh 2012: 1122).

This begs another question: What is the best approach to inform such deliberative participatory mechanisms, and thus better community engagement? In the participatory literature, collaborative approaches are best seen and evidenced to engender trustful and insightful engagement (e.g. Ellis et al. 2006). Collaborative process in the deliberative sense invites early involvement in decision-making process, transparency of information and process, inclusiveness, intersubjective dialogue, participant diversity, broad representation and partnership in the agenda setting of the process and its goals (e.g. Edwards et al. 2008, Hindmarsh 2010, Rowe and Frewer 2000). This approach contrasts to the traditional top-down and shallower 'inform-consult-involve' spectrum of community involvement (e.g. IAP2 2007), through, for example, inquiry or planning submissions, town hall meetings, or information sessions where community decisional influence is distinctly lacking (Wolsink 2007). In energy management, such approaches have increasingly emerged since the late 1990s (e.g. Cotton and Devine-Wright 2010, Ellis et al. 2006, Juraku et al. 2007, O'Connor and van den Hove 2001). Of particular importance to collaborative approaches is to motivate citizens to assume place 'co-ownership' of local management problems in addition to contributing to generating and implementing solutions (Beierle and Konisky 1999, Bidwell and Ryan 2006).

Concomitantly, well-worn criticisms of inclusive participatory approaches include the time and resources needed to build participant diversity and representativeness; how interested, informed and educated the public is for participation; whether too much participation will lead to overly reactionary responses that lead to confrontation and ineffective policy results; and the appropriateness of processes and mechanisms for participation to influence policy outcomes. However, with the benefits of collaboration convincing, Jones and Eiser (2010: 3116) referred to '[a] burgeoning literature [that] now exists to attest to the many benefits of employing more inclusive and deliberative approaches to project development . . . [that] firmly points to the importance of early, sustained and reciprocal interactions'.

Returning to my emphasis on place-change, this emphasis also reflects increasing international recognition of the relevance of incorporating 'new localism', 'place', and 'community' into public policy and planning 'in part due to . . . the deficiencies of traditional centralised bureaucratic arrangements' to achieve legitimate and effective planning decisions (Eversole 2011: 52). This recognition marks decades of increasing public distrust and social conflict at the local place-based community level with regard to controversial facility siting, as well as environmental management and planning policy or practices. Such conflict involving strong local contestation of plans that are seen to discount local views too much is informed increasingly by science, technology and environmental citizenship and democratization movements, and trends toward strong sustainability and climate change adaptation whole-of-society approaches (Connelly et al. 2011, Dovers 2005). Special attention is paid to meaningful or 'active' public participation (e.g. Hindmarsh 2010, Van der Horst 2007), like the citizen referendum that Juraku et al. (2007) reported on.

In this terrain, place contestation to facility siting is then not seen simplistically as nimbyism or outright opposition for the sake of it (Wolsink 2007), but instead as 'place-protectionism' to 'place disruption' (Devine-Wright 2009). 'Place' is thus the underlying value reference point for place-protectionism. Traditional policy and planning approaches are not informed by the concept of place but instead of 'community', so miss the value or point of exploring 'place' to make better sense of why there is social conflict around plans for, or embedment of, controversial facility siting in place, especially involving megatechnologies. In addition, there is little interest from top-down systems to probe deeper to better understand social conflict at the local level. As a result, simplistic understandings of the issues posed at the place-based community level abound, and the result most often is ill-conceived plans, which, again, is one important reason why the Fukushima Daiichi disaster occurred (see also Falk in this book).

'PLACE' AND 'PLACE-CHANGE'

But in developing place and place-change understandings of local communities in cultural and practice context, what exactly is a place-based local community? Reinforcing the notion of 'community diversity' as found in most landscapes, for Colclough and Sitaraman (2005: 478), a 'place-based local community' is a subset of all possible community types 'best viewed as communities in place versus communities of place'. Thus, it is not the common experience of place that gives rise to these communities but the common experience that occurs *within* place. While displaying shared belonging and attachment to place, such communities (and their members) can also be quite diverse in experiences of place (also Shucksmith 2010). This is because of social relationships, the complexity of social organization (Colclough and Sitaraman 2005: 478), interests, 'and the particular biophysical environment or landscape that embeds, binds or bounds them' (Hindmarsh 2012: 1130). Regarding sociospatial parameters, rural place-based communities are typically found 'within a defined area . . . interacting and participating in a wide range of local affairs, and sharing an awareness of common life and personal bonds' (Dalton and Dalton 1975: 2). Such diversity of local place-based communities highlights the point of Juraku et al. (2007) that improvements or revisions of existing decision-making systems on facility sting need to build in flexibility, 'to adapt to the context and situation of each case'.

Accordingly, (each) 'place' represents a complex psychological construction of social and environmental attributes and meanings, assigned by individuals by way of sociocultural processes (Pretty et al. 2003). It forms an underlying value reference point (or rationality context) for the formation of local beliefs and attitudes to place and to place-change and disruption (Woolley 2010), including the radical change associated with the siting of megatechnologies like nuclear power plants. The reference point of place is often referred to as 'place attachment' (e.g. Devine-Wright 2009) or 'sense of place' (Nadaï 2007). Devine-Wright (2009: 427) defines place attachment as 'both the process of attaching oneself to a place and as a product of this process'. Such attachment is influenced by positive emotional connections with familiar locations such as the home or neighborhood. It correlates with length of dwelling, and features many sociobiophysical subdimensions. Alternatively, for Nadaï (2007), sense of place refers to local social connections to the encompassing landscape that holds it together as a unified and distinctive entity. Similarly, Hummon (1992: 262) found sense of place fuses one's understandings of, and feelings about, place. Positive place attachment was, in turn, found to contribute to an individual's sense of self or identity, personally and socially (Devine-Wright 2009: 428). Such identity facilitates a person's lifestyle in reflecting both held and formative values and norms (Relph 1976: 43).

These values and norms, in reflecting positive and negative experiences associated with a place, also inform one's sense of 'right' and 'wrong', or what is acceptable or not in the world and in or to 'place' (Proshansky et al. 1983). Acceptability to nuclear power plant siting, for example, then relates intimately to any place-change, proposed or imposed. This is especially the case when such change is seen to affect a person's self-identity and emotional well-being (Wester-Herber 2004: 114), and particularly in relation to cherished environmental and social characteristics of place (Van der Horst 2007). Anxiety, grief, depression and a sense of loss to individuals can result if adverse or unacceptable change occurs, for example, with regard to the dread, stigmatization and stress often felt by communities hosting nuclear facilities. Such feelings are enhanced if such change occurs through an unjust manner such as community engagement lacking fair and adequate processes of local inclusion in decision making (Wolsink 2007: 1202–1204).

Place and place-change understandings—and the 'liberation' of in-depth social knowledges relevant to the problem at hand through understandings gained through collaborative participatory mechanisms—can then offer valuable insights for better local planning, especially pertinent for radical change posed by megatechnologies. These understandings include the varying degrees of significance of place for individuals, groups and communities, and other local place siting qualifications (or boundaries), particularly about the various physical and symbolic attributes that contribute to sense of self or identity. In addition, such insights help to identify what poses as place-disruption of a radical nature and how to better mitigate it with regard to place-protectionist needs and priorities; something that Japanese communities were unable to realize in any meaningful way in the long history of nuclear build in Japan.

CONCLUSION

The Fukushima Daiichi disaster poses many implications for change. The implication for change posed in this chapter relates to the important policy question of the inadequacy of public participation in decision making about nuclear power plant siting in Japan as a key problem contributing to the Fukushima Daiichi disaster. It explored this implication in science, technology and society contexts of science, technology and governance, and sociotechnical systems. It focused on this topic particularly in relation to the facility siting in local place-based communities of nuclear megatechnological assemblages that pose high risk of 'radical' change, which the Fukushima Daiichi disaster put under the spotlight worldwide.

A key policy envelope framing the chapter's inquiry was the notion of 'good governance' relating to long-term social and environmental sustainability, here, in relation to energy and also health (both human and nonhuman), as informed by principles of openness, participation, accountability, effectiveness and coherence (Commission of the European Communities 2001). Three research questions were posed: How did the sociopolitical dynamics of nuclear power development and regulation and inadequate public engagement about nuclear power, especially the siting of power plants, in Japan contribute to unsustainable energy outcomes (as finally revealed by the nuclear disaster)? How was nuclear power development and regulation constructed and maintained to keep public engagement weak? Finally, what change might be suggested to contribute to less risk of disaster and more scope for sustainable energy futures?

In addressing these questions, another key context shaping the scope of the Fukushima Daiichi disaster—as a megadisaster involving an unprecedented three-reactor meltdown—was the nature of its substance. On one hand were the natural components of the preceding earthquake and tsunami; on the other hand, the social components that marked it also as a 'chronic technological disaster' involving social actors, decisions and policies or lack thereof found in the discursive interplay of the various stakeholders involved and their practices and actions, the synergy of which (inadvertently) invited high vulnerability to natural disaster (see Gramling and Krogman 1997: 42). This casual duality or coproduction of the disaster was well evidenced by the dramatic swing in public opinion after the disaster to one strongly of denuclearization, in reaction to the nuclear disaster itself and to the escalating lack of trust in the government and the nuclear industry to adequately ensure nuclear safety and more specifically to manage the disaster, with the Tokyo Electric Power Company the industry culprit.

As such, in respect to the first two questions, the analysis points to inadequate community engagement as a key social aspect informing mismanagement of the siting of nuclear power plants in Japan as biased more toward a technocratic approach based on technical criteria than a partnership one of both technical and social criteria. This was despite these megatechnological nuclear power assemblages posing many diverse system uncertainties and complexities, of both a social and biophysical/geophysical nature in relation to their safety and site selection. Such practice thus reflects a flawed planning system that had little connection to the social preferences and knowledge of the local place-based communities into which these risky technological assemblages were inserted. If local social knowledges and preferences had been included, very likely, siting locations and pathways across Japan may well have been quite different and averted multisiting in highly dangerous places, as Fukushima again demonstrated. Local knowledge exclusion well indicated a break down in the Japanese sociotechnical system of nuclear power development, which heavily contributed to a dysfunctional siting and development program. In short, the *centralized* governed technical infrastructure clearly became disconnected from its *decentralized* (and 'democratically conditioned') social infrastructure of rural local communities across Japan through marginalizing local voices over a long time about siting issues.

Instead of engaging with communities in well-thought-out collaborative engagement strategies to best draw on local perspectives and social knowledge about place as a facility site, Japanese nuclear authorities and industries chose the path of least resistance to nuclear progress. Creating and then nurturing compliant nuclear villages was chosen on the basis of lack of resistance rather than on the safest environment, that is, with regard to siting as more conditioned by less than more nuclear reactors and location. The outcome of this flawed planning approach saw the advent of serial multiple siting of reactors in one place, as well as clusters of concentrated siting plants, as demonstrated in the Fukushima prefecture. Such siting policy and planning practices eventually and perhaps inevitably enabled megadisaster (see also Juraku in this book).

These findings then informed how to address the third and final question: What change might be suggested to contribute to less risk of disaster and more scope for sustainable energy futures? With regard to the focus on community engagement in this chapter', and in agreement with Japanese STS researchers Juraku et al. (2007), a modification of existing planning systems through the insertion of civic deliberative participatory spaces appears the best way forward. Such dialogic spaces can best enable meaningful engagement of local communities for their input into facility siting processes to engender more legitimate and effective planning outcomes (also Lidskog

2005). Accordingly, this chapter has advanced the concept of 'place-change planning' as a facilitative conduit to better engage diverse local place-based communities with planners and developers through deliberative spaces, arenas and pathways for more plural, social situated understandings, and local knowledge input, about where, when and what to site in targeted situations (also Bergmans 2008, Selman 2004, Stirling 2008).

But for any change involving alternatives to nuclear, such as renewable energy, it should also be noted that the same arguments apply to their adoption, where wind energy has so far in the renewable energy portfolio best illustrated the case for enhanced community engagement. Many of the same problems of significant disconnect between technical and social infrastructures, as shown in nuclear facility siting, characterize wind farm facility siting. Incidentally, this has tended to hamper effective renewable energy transitions through increasing local social conflict, lack of legitimacy and poor planning (e.g. Hindmarsh 2010, van der Horst 2007, Wolsink 2007). Indeed, exploring inadequate community engagement around the many local issues of wind energy systems being introduced in Australia, and in global context, offered the primary source of evidences to conceptualize place-change planning by which to, in this case, apply to Japan's nuclear energy siting issues.

What is then highlighted is that the real problem in developing integrated whole-of-society sociotechnical system functionality may not be so much about the choice of technology—apart from any particular problems of unpredictability and complexity and benefit-claim and specific issues that each megatechnology poses. Instead, it seems apparent that it is first and foremost one of successfully bridging the ambivalent relationships and tensions between democracy and technology, the politics of place, expert/lay and social/technical divides, and development and sustainability. Indeed, such bridging reflects a core implication and lesson of the tragedy of the nuclear disaster at Fukushima Daiichi power plant of the urgent need to shift to 'democratic technology', or to democratize technology, (more broadly, technoscience) according to principles of good governance and strong sustainability. Such a shift would explicitly support enhanced community engagement—along the lines of place-change planning, for example—for any siting planned or proposed of megatechnologies that pose radical change for local place-based communities, in this case, that of the nuclear one.

ACKNOWLEDGMENTS

The author would like to acknowledge the support of the Australian Research Council Discovery Projects Scheme (project DP0986201), and the Centre for Governance and Public Policy, Griffith University, and contribution to research by Anne Parkinson.

REFERENCES

Aldrich, D. "Location, Location, Location: Selecting Sites for Controversial Facilities." *Singapore Economic Review* 53, no. 1 (2008): 145–172.

Asahi Shimbun. "Interview/Ulrich Beck: System of Organized Irresponsibility behind the Fukushima Crisis." *Asahi Shimbun,* July 6, 2011, accessed October 27, 2012. http://ajw.asahi.com/article/0311disaster/opinion/AJ201107063167.

Avenell, S. "From Fearsome Pollution to Fukushima: Environmental Activism and the Nuclear Blind Spot in Contemporary Japan." *Environmental History* 17 (2012): 244–276.

Beck, U. *Risk Society: Towards a New Modernity.* London: Routledge, 1992.

Beck, U. *Ecological Enlightenment: Essays on the Politics of the Risk Society.* Atlantic Highlands, NJ: Humanities Press, 1995.

Beck, U. "Politics of Risk Society." In *The Politics of Risk Society,* edited by J. Franklin, 9–22. Cambridge: Polity Press, 1998.

Beierle, T., and D. Konisky. *Public Participation in Environmental Planning in the Great Lakes Region.* DP99–50. Washington, DC: Resources for the Future, 1999.

Bergmans, A. "Meaningful Communication among Experts and Affected Citizens on Risk: Challenge or Impossibility?" *Journal of Risk Research* 11, no. 1 (2008): 175–193.

Bidwell, R., and C. Ryan "Collaborative Partnership Design: The Implications of Organizational Affiliation for Watershed Partnerships." *Society and Natural Resources* 19, no. 9 (2006): 827–843.

Bijker W. *Of Bicycles, Bakelites, and Bulbs: Toward a Theory of Sociotechnical Change.* Cambridge, MA: MIT Press, 1995.

Bishop, R., and D. Vogel. "Power Plant Siting on Wisconsin's Coasts: A Case Study of a Displaceable Use." *Coastal Zone Management Journal* 3, no. 4 (1977): 363–384.

Blowers, A. "Why Dump on Us? Power, Pragmatism and the Periphery in the Siting of New Nuclear Reactors in the UK." *Journal of Integrative Environmental Sciences* 7, no. 3 (2010): 157–173.

Butler, D. "Reactors, Residents and Risk." *Nature,* April 21, 2011, accessed October 25, 2012. http://www.nature.com/news/2011/110421/full/472400a.html.

Cabinet Office. " 'Genshi-ryoku ni Kansuru Tokubetsu Seron Chousa' no Gaiyou [Overview of 'The Special Poll on Nuclear Power']." November 26, 2009, accessed May 21, 2012. http://www8.cao.go.jp/survey/tokubetu/h21/h21-genshi.pdf, cited in Hara in this book.

Cochran, T., and M. McKinzie. *Global Implications of the Fukushima Disaster for Nuclear Power.* Natural Resources Defense Council, 2012.

Colclough, G., and B. Sitaraman. "Community and Social Capital: What Is the Difference?" *Sociological Inquiry* 75, no. 4 (2005): 474–496.

Commission of the European Communities (CEC). *European Governance: A White Paper.* Com (2001): 428. Brussels: CEC. 2001.

Connelly, S., S. Markey, and M. Roseland. "Bridging Sustainability and the Social Economy: Achieving Community Transformation through Local Food Initiatives." *Critical Social Policy* 31 (2011): 308–324.

Connor, C. "A Quantitative Literacy View of Natural Disasters and Nuclear Facilities." *Numeracy* 4, no. 2 (2011): Article 2. DOI: 10.5038/1936–4660.4.2.2.

Corner, A., D. Venables., A Spence., W. Poortinga., C. Demski, and N. Pidgeon. "Nuclear Power, Climate Change and Energy Security: Exploring British Public Attitudes." *Energy Policy* 39 (2011): 4823–4833.

Cotton, M., and P. Devine-Wright. "Making Electricity Networks 'Visible': Industry Actor Representations of 'Publics' and Public Engagement in Infrastructure Planning." *Public Understanding of Science* 17 (2010): 1–19.

Dalton and Dalton. "Community and Its Relevance to Australian Society: An Examination of the Sociological Definition." Prepared for the Department of Tourism and Recreation by Dalton and Dalton. Canberra: Australian Government Publishing Service, 1975.
Devine-Wright, P. "Rethinking NIMBYism: The Role of Place Attachment and Place Identity in Explaining Place-Protection Action." *Journal of Community and Applied Social Psychology* 19, no. 6 (2009): 426–441.
Dovers, S. *Environment and Sustainability Policy: Creation, Implementation, Evaluation.* Sydney: Federation Press, 2005.
Du Plessis, R., R. Hindmarsh, and K. Cronin. "Engaging across Boundaries—Emerging Practices in 'Technical Democracy.'" *East Asian Science, Technology and Society: An International Journal* 14 no. 4 (2010): 475–482.
Edwards, P., R. Hindmarsh, H. Mercer, M. Bond, and A. Rowland. "A Three-Stage Evaluation of a Deliberative Event on Climate Change and Transforming Energy." *Journal of Public Deliberation* 4, no. 1 (2008): Article 6, accessed October 29, 2012. http://services.bepress.com/jpd/vol4/iss1/art6.
Ellis, G., J. Barry, and C. Robinson. *Renewable Energy and Discourses of Objection: Towards Deliberative Policy-Making: Summary of Main Findings.* Northern Ireland: Queen's University Belfast, 2006.
Eversole, R. "Community Agency and Community Engagement: Re-theorising Participation in Governance." *Journal of Public Policy* 31, no. 1 (2011): 51–71.
Fischer, F. *Technocracy and the Politics of Expertise.* London: Sage, 1990.
Fischer, F. "Technological Deliberation in a Democratic Society: The Case for Participatory Inquiry." *Science and Public Policy* 26, no. 5 (1999): 294–302.
Freudenburg, W. "Contamination, Corrosion and the Social Order: An Overview." *Current Sociology* 45, no. 3 (1997): 19–39.
Fujigaki, Y., and T. Tsukahara. "STS Implications of Japan's 3/11 Crisis." *East Asian Science, Technology and Society: An International Journal* 5 (2011): 381–394.
Gramling, R., and N. Krogman. "Communities, Policy and Chronic Technological Disasters." *Current Sociology* 45, no. 3 (1997): 41–57.
Greenberg, M., F. Popper, and H. Truelove. "LULUs Still Enduringly Objectionable?" *Journal of Environmental Planning and Management* 55, no. 6 (2012): 1–19.
Gregory, R., and T. Satterfield. "Beyond Perception: The Experience of Risk and Stigma in Community Contexts." *Risk Analysis* 22, no. 2 (2002): 347–358.
Haraway D. *Modest_Witness@Second_Millennium.FemaleMan©_Meets_OncoMouse: Feminism and Technoscience.* New York: Routledge, 1997.
Hindmarsh, R. "Wind Farms and Community Engagement in Australia: A Critical Analysis for Policy Learning." *East Asian Science, Technology and Society: An International Journal* 4, no. 4 (2010): 541–563.
Hindmarsh, R. "'Liberating' Social Knowledges for Water Management, and More Broadly Environmental Management, through 'Place-Change Planning.'" *Local Environment: The International Journal of Justice and Sustainability* 17, no. 10 (2012): 1121–1136.
Hummon, D. "Community Attachment: Local Sentiment and Sense of Place. In *Place Attachment,* edited by I. Altman and S. Low, 253–278. New York: Plenum Press, 1992.
IAP2 (The International Association for Public Participation). "IAP2 Spectrum of Participation," 2007, accessed June 11, 2010. http://www.iap2.org/associations/4748/files/IAP2%20Spectrum_vertical.pdf.
Jones, C., and R. Eiser. "Understanding 'Local' Opposition to Wind Development in the UK: How Big Is a Backyard?" *Energy Policy* 38 (2010): 3106–3117.
Juraku, K., T. Suzuki, and O. Sakura. "Social Decision-Making Processes in Local Contetxs: An STS Case Study on Nuclear Power Plant Siting in Japan." *East Asian Science, Technology and Society: An International Journal* 1 (2007): 53–75.

Kikkawa, T. *Dynamism of Development in the Japanese Electric Power Industry*. Nagoya: Nagoya University Press, 2004.
Kunreuther, H., D. Easterling, W. Desvousges, and P. Slovic. "Public Attitudes towards Siting a High Level Nuclear Waste Repository in Nevada." *Risk Analysis* 10 (1990): 469–484.
Latour, B. *Science in Action: How to Follow Scientists and Engineers through Society*. Cambridge, MA: Harvard University Press, 1987.
Lidskog, R. "Siting Conflicts–Democratic Perspectives and Political Implications." *Journal of Risk Research* 8, no. 3 (2005) 187–206.
Matanle, P. "The Great East Japan Earthquake, Tsunami, and Nuclear Meltdown: Towards the (Re)construction of a Safe, Sustainable, and Compassionate Society in Japan's Shrinking Regions." *Local Environment: The International Journal of Justice and Sustainability* 9 (2011): 823–847.
Miller, C., D. Sarewitz, and A. Laight. *Science, Technology, and Sustainability: Building a Research Agenda*. Report for the National Science Foundation, USA, 2008.
Nadaï, A. " 'Planning,' 'Siting' and the Local Acceptance of Wind Power: Some Lessons from the French Case." *Energy Policy* 35 (2007): 2715–2726.
Nahuis, R., and H. van Lente. "Where Are the Politics? Perspectives on Democracy and Technology." *Science, Technology & Human Values* 33 (2008): 559–581.
Nakamura, A., and M. Kikuchi. "What We Know, and What We Have Not Yet Learned: Triple Disasters and the Fukushima Nuclear Fiasco in Japan." *Public Administration Review,* November/December 2011, 893–899.
O'Connor, M., and S. van den Hove. "Prospects for Public Participation on Nuclear Risks and Policy Options: Innovations in Governance Practices for Sustainable Development in the European Union." *Journal of Hazardous Materials* 86, nos. 1–3 (2001): 77–99.
Owens, S. "Siting, Sustainable Development and Social Priorities." *Journal of Risk Research* 7, no. 2 (2004): 101–114.
Perrow, C. *Normal Accidents: Living with High-risk Technologies*. New York, London: Basic Books, 1984.
Pidgeon, N., I. Lorenzoni, and W. Poortinga. "Climate Change or Nuclear Power–No Thanks! A Quantitative Study of Public Perceptions and Risk Framing in Britain." *Global Environmental Change* 18 (2008): 69–85.
Pretty, G., H. Chipuer, and P. Bramston. "Sense of Place amongst Adolescents and Adults in Two Rural Australian Towns." *Journal of Environmental Psychology* 23 (2003): 273–287.
Proshansky, H., A. Fabian, and R. Kaminoff. "Place-Identity: Physical World Socialization of the Self." *Journal of Environmental Psychology* 3 (1983): 57–83.
Relph, E. *Place and Placelessness*. London: Pion, 1976.
Rowe, G., and L. Frewer. "Public Participation Methods: A Framework for Evaluation." *Science, Technology & Human Values* 25 (2000): 3–29.
Selman, P. "Community Participation in the Planning and Management of Cultural Landscapes." *Journal of Environmental Planning and Management* 47, no. 3 (2004): 365–392.
Shucksmith, M. "Disintegrated Rural Development? Neo-endogenous Rural Development, Planning and Place-shaping in Diffused Power Contexts." *Sociologia Ruralis* 50, no. 1 (2010): 1–14.
Spink, A., M. Hillman, K. Fryis, G. Brierly, and K. Lloyd. "Has River Rehabilitation Begun? Social Perspectives from the Upper Hunter Catchment, New South Wales, Australia." *Geoforum* 41 (2010): 399–409.
Starr, S. "The Ethnography of Infrastructure." *American Behavioral Scientist* 43, no. 3 (1999): 377–391.

Stirling, A. " 'Opening Up' and 'Closing Down': Power, Participation, and Pluralism in the Social Appraisal of Technology." *Science, Technology & Human Values* 33 (2008): 262–294.

Tamman, M., B. Casselman, and P. Mozur. "Scores of Reactors in Quake Zones." *Wall Street Journal,* March 19, 2011, accessed November 20, 2011. http://online.wsj.com/article/SB10001424052748703512404576208872161503008.html.

Tanaka, Y. "Major Psychological Factors Determining Public Acceptance of the Siting of Nuclear Facilities." *Journal of Applied Social Psychology* 34, no. 6 (2004): 1147–1165.

Turiel, E. *The Development of Social Knowledge: Morality and Convention.* Cambridge: Cambridge University Press, 1983.

Uekoetter, F. "Fukushima, Europe, and the Authoritarian Nature of Nuclear Technology." *Environmental History* 17 (2012): 277–284.

Unger, S. "Moral Panic versus the Risk Society: The Implications of the Changing Sites of Social Anxiety." *British Journal of Sociology* 52, no. 2 (2001): 271–291.

Urbanik, J. "Locating the Trangenic Landscape: Animal Biotechnology and Politics of Place in Massachusetts." *Geoforum* 38 (2007): 1205–1218.

Vanclay, F. "Place Matters." In *Making Sense of Place*, edited by F. Vanclay, M. Higgins and A. Blackshaw, 3–11. Canberra: National Museum of Australia Press, 2008.

Van der Horst, D. "NIMBY or Not? Exploring the Relevance of Location and the Politics of Voiced Opinions in Renewable Energy Siting Controversies." *Energy Policy* 35 (2007): 2705–2714.

Wester-Herber, M. "Underlying Concerns in Land-use Conflicts: The Role of Place-identity in Risk Perception." *Environmental Science and Policy* 7 (2004): 109–116.

Wolsink, M. "Planning of Renewables Schemes: Deliberative and Fair Decision-making on Landscapes Issues Instead of Reproachful Accusations of Non-cooperation." *Energy Policy* 35, no. 5 (2007): 2692–2704.

Woolley, O. "Trouble on the Horizon: Addressing Place-Based Values in Planning for Offshore Wind Energy." *Journal of Environmental Law* 22, no. 2 (2010): 223–250.

Yamaguchi, M. "Japan's Cabinet Backpedals on Totally Phasing out Nuclear Energy; New Regulatory Panel Starts." *Canadian Press*, September 19, 2012, accessed November 1, 2012. http://essentialforbody.blogspot.com.au/2012/09/japans-cabinet-backpedals-on-totally.htm.

Yoshioka, H. *Shin-ban Genshi-ryoku no Shakai-shi: Sono Nihon-teki-tenkai [Social History of Nuclear Power: The Trajectory in Japan],* 2nd ed. Tokyo: Asahi–'shimbun Shuppan, 2011, cited in Hara in this book.

5 Environmental Infrastructures of Emergency

The Formation of a Civic Radiation Monitoring Map during the Fukushima Disaster

Atsuro Morita, Anders Blok, and Shuhei Kimura

INTRODUCTION: CIVIC INFRASTRUCTURES OF EMERGENCY

This chapter discusses the role of civic engagement in the measurement of atmospheric radiation levels, and the stabilization of an emergent measurement infrastructure, during the immediate state of emergency in the early days following the Fukushima Daiichi disaster on March 11, 2011. Focusing on the specific case of a 'civic radiation monitoring map' voluntarily organized through ad hoc collaboration among a group of 'amateur' Japanese citizens from March 13 onward, we discuss the importance, in a situation of widespread public unrest and mistrust in official institutions, of civic efforts to make radiation data publicly accessible and more easy to understand (see also Kera et al. in this book). Civic engagement with what was hitherto a mostly hidden environmental information system—that of networked radiation monitoring posts—turned out to be crucial, we argue, for filling one piece of the serious void of technoscientific credibility left open within Japan by the Fukushima Daiichi disaster. In this sense, the resultant monitoring map, which we refer to as a civic infrastructure, turned out to be an appropriate technological and political response to an emergency situation of extreme uncertainty and lack of reliable information.

In analytical terms, this ad hoc civic radiation measurement collaboration can be considered to be a case of what Michel Callon and others (e.g. Callon and Rabeharisoa 2008) call 'emergent concerned groups'. While this notion belongs to the family of pragmatist political philosophy elaborated in recent work by actor network theorists (ANT) within science, technology and society (STS) studies (e.g. Latour 2007, Marres 2007), we deploy it here mainly for its own 'intuitive' (or heuristic) value, without invoking the full methodology of ANT.[1] What is particularly valuable, for our purposes, in the notion of emergent concerned groups is that it highlights how public engagement with technoscience is called for at exactly those moments when established institutional frameworks fail, collapse, or otherwise prove inadequate in addressing collectively experienced problems. As such, it enables us to take a particular analytical approach to wider STS debates on the importance of civic participation in handling technoscientific risks, such as the radiation risks of nuclear power.

Seen from within Japan immediately after the Fukushima Daiichi disaster, institutional frameworks and responses were clearly seen to fail, collapse, or otherwise prove inadequate. Evoking Callon's suggestive concepts (1998), the days following the Fukushima Daiichi disaster is a paradigmatic case of a 'hot'—indeed, 'super-hot'—situation, characterized by technoscientific controversy, disorientation, and a variety of material and symbolic uncertainties (or 'overflows'). In such situations, ad hoc publics come into being, aiming to address their collective sense of concern through coordinating their own forms of civic inquiry (cf. Callon and Rabeharisoa 2008). By analyzing the concrete activities of concerned Japanese citizens, as they assemble and collectively establish a civic radiation monitoring map, we aim to contribute to the wider objectives of this book (see Chapter 1 in this book for detail), in terms of understanding the technological and political importance of new citizen networking practices related to the contentious issue of nuclear risks in Japan.

In what follows, we describe how a small group of Japanese citizens helped assemble a radiation monitoring infrastructure that was more reliable, more credible, more locally responsive—in short, more 'appropriate'—than what otherwise existed. Following Fortun (2004), we suggest that this emerging civic environmental infrastructure was an *appropriate* technology in the sense that it was attuned to the material, political, and technological realities within which it worked, and to the people who used it. In the immediate aftermath of the Fukushima Daiichi disaster, these realities included not just an acute sense of technoscientifically induced danger, but also a serious lack of credible and actionable knowledge, given the widespread mistrust in information released by the Japanese government. In this sense, the civic radiation measurement infrastructure, like all infrastructure (Star 1999), is relational; it is driven largely by the *breakdown* of other related infrastructures.

In the wake of the Fukushima Daiichi disaster, Japanese bureaucratic practices gave rise to formulaic government announcements, uniformly reported by mainstream television channels and newspapers that would repeatedly emphasize the limited influence of radiation on human health—without, however, providing any details. On the other hand, the very climate of suspicion also meant that public spaces came to be saturated by a variety of speculations, hearsays, rumors, and alleged confidential information. These appeared on Twitter accounts, on web blogs, and in foreign media sources, often alluding to a 'hidden reality' of catastrophic events ongoing at the plant. From the point of view of an ordinary Japanese citizen, realities indeed multiplied, as one plunged through ever-escalating media-borne controversies.

In this situation of extreme uncertainty, the civic radiation monitoring map, assembled by citizens using freely available software (Google Maps), came to play an important role as a tool to support people's assessments and actions. Like other environmental information infrastructures, the civic map gradually came to depict what people 'saw' in their immediate surroundings (Fortun 2004). The map provided a means not only to know the real-time radiation levels in (or near) one's locality, but also to assess the plausibility and credibility of ('reassuring') official press releases and ('catastrophic') circulating rumors. While contested by certain specialists on technical grounds, the emerging citizen-driven network of radiation monitoring proved *practically* sufficient for people to know and evaluate the fluctuating tendencies of radiation levels. The civic map became an important 'cartographic' tool to help citizens somewhat regain their *grip* on an otherwise fractured reality.

Ultimately, we suggest that the sense of credibility with which the map became invested was made possible by its very *heterogeneous* sociotechnical constitution, that is, by linking available data regardless of the social backgrounds or levels of technical proficiency of the people doing the monitoring. Hence, importantly, data represented on the map came from monitoring posts run by a variety of entities. These included citizens, local municipalities, power companies, universities, and other research institutions. The interplay of such diverse measurements could then be taken to significantly lessen political 'bias', thereby constituting the closest one could get, in the situation, to some sense of 'objectivity' (cf. Daston and Galison 2007).

This observation resonates with the claim of STS theorist Susan Leigh Star (e.g. 1999) that the heterogeneous quality of infrastructures—embedded in divergent protocols and contexts of usage—is integral to their proper functioning, in terms of reproducing ordinary versions of reality without being noticed as doing so. In our case, this functioning was, in turn, made publicly reflexive through the map as a tool of self-reference informing citizen assessment and action. In other words, the map performed what Bowker (1994) calls an 'infrastructural inversion', which made visible a hitherto back-grounded infrastructure of government- and industry-run radiation monitoring. In making publicly visible the inner workings of this infrastructure, the 'newly added' citizen-run measurement posts also served, perhaps inadvertently, to gradually reestablish trust in more official data sources.

In what follows, we start by tracing the contours of how public mistrust in official radiation statements was shaped by particular 'un-civic' forms of technoscientific practices of the Japanese government and mainstream media (cf. Jasanoff 2005), and in turn set the scene for emergent concerned citizen groups of radiation monitoring. We then trace the gradual constitution of these groups, together with their main infrastructural reference point, the radiation monitoring map, as a 'sociotechnical network'. Upon outlining how the map performed its infrastructural inversion, and why this was important for restoring some sense of public grip on the situation, we conclude with a more general discussion of what this particular case, in the wake of a Japanese nuclear disaster, tells us about the role of public engagement with 'risky' technosciences.[2] Here, we revisit the pragmatist lineage of political philosophy—committed to notions of experimentation, public intelligence and collective learning—as a way of making sense of post-Fukushima responses by the Japanese public.

In terms of methods, our discussion builds primarily on data, collected through virtual (or online) ethnography, on the day-to-day dynamics of the radiation monitoring map and the concerned citizens co-constituting it.[3] As such, primary data consisted of screen captures and copies of web pages created and downloaded in the midst of the ongoing events. When making broader claims about the Japanese public at large, we rely also on auto-ethnography, given that we (this chapter's authors) were all living through the Fukushima Daiichi disaster from within a Japanese setting (Kyoto/Osaka/Shizuoka), with full exposure to various media sources, public spaces, and daily conversations with family, friends, and colleagues.[4] While we thus want to acknowledge the tentative and situated character of our claims, we believe this situation mirrors closely our position as 'coincidental ethnographers' caught up in dramatic sociotechnical events.[5] If what we write about is emerging concerned groups in a situation of emergency, there is a sense in which our own writing is also an emerging concerned scholarly reflection on unsettling events that, evidently, are yet to be fully settled, in and beyond official politics.[6] We hope our reflection may itself become a small piece of this process.

THE MAKING OF PUBLIC MISTRUST: OFFICIAL 'UN-CIVIC' EPISTEMOLOGY?

In her book *Designs on Nature* (2005: chapter 10), STS scholar Sheila Jasanoff coins the term 'civic epistemology' in reference to the institutionalized practices by way of which different political cultures test the public credibility of technoscience. Civic epistemologies include, for instance, shared notions of what constitutes a valid public demonstration of proof; what sources of expertise should be considered legitimate; and on what basis members of the public are supposed to trust their authorities in matters of technoscientific politics. As important parts of how national political cultures vary, civic epistemologies are crucial for understanding how specific societies deploy knowledge claims as a basis for making collective choices. Similarly, civic epistemologies are an important lens through which to understand public engagements with technoscience, in terms of how acceptance or rejection of particular technologies, like nuclear power, is historically and culturally shaped.

Jasanoff's original work covered the US, UK, and Germany. In this context, part of what is noticeable about the Fukushima Daiichi disaster, we suggest, is that it allows the world, and the Japanese public, to get a glimpse of how official *Japanese* civic epistemology works—or, indeed, does *not* work—under extreme pressures of emergency. In the days following the disaster, a deep crisis of credibility opened up around official Japanese institutions of government, industry and media, exposing to public view their rather 'un-civic' ways of handling nuclear threats. This failed official performance of crisis management, and the widespread public mistrust it bred within Japan, is a crucial factor in understanding the emergence of new radiation-monitoring publics. To recount this story, we first need to look carefully at the day-to-day unfolding of official announcements on the disaster, and how these were met with public suspicion and mistrust.

From the point of view of the Japanese public, the early days of the Fukushima Daiichi disaster were characterized not only by lack of reliable information, but by a flooding of various discourses, via mainstream media and through online forums, telling starkly different versions of the unfolding events (Ogiue 2011). Accounts were publicized, through different channels, by a variety of parties: spokespersons of the Japanese government; major media corporations; antinuclear groups; freelance journalists; foreign media; and various experts from home and abroad. This cacophony of voices entailed a perplexing feeling of multiplying realities, as the Fukushima Daiichi nuclear power plant was transformed, almost in an instant, from a little-known technological object into a global and hugely controversial, evolving, and uncertain 'matter-of-concern'.[7] Like other members of the public, we (the authors) became serially immersed in vastly different versions of the disaster. As new story lines, information bits, and rumors came onto the scene every hour, it felt a bit like constantly slipping from one world into another.

We may recall this sensation by (selectively) reviewing how stories multiplied and diverged from each other in the series of events that followed the first official announcement of the 'incident' on March 11, 2011. According to official press releases, Tokyo Electric Power Company (TEPCO) made its first report to the government on the state of total electricity deprivation at reactors 1, 2, and 3 at half past four in the afternoon, two hours after the earthquake. This implied the loss of means to cool down the stopped reactors. Two and a half hours later, chief secretary of the Japanese cabinet, Yukio Edano, declared a 'state of nuclear emergency' during a press conference. This marked the first public notification of the disaster. Edano's announcement became more detailed later that evening, as first the Fukushima governor and then Prime Minister Naoto Kan announced their directives of evacuation from the area within a three-kilometer radius from the plant.[8] Still, at this point in time, most citizens probably did not sense the seriousness of the situation, as government announcements gave the impression of a 'minor incident', while mainstream media reports repeatedly emphasized that things were safe and under control.

Public trust in the official story line of safety was seriously shaken by successive events starting the next day, on March 12. At 3 p.m. that day, the building housing reactor 1 suddenly exploded. While the government was quick to announce that the explosion had occurred in the space outside the reactor, emphasizing that the reactor itself was safe, official press releases about this unexpected event cultivated citizen mistrust. Fear of radioactive emissions from the reactors started to materialize, due to a steady increase in air radiation dose rates measured by existing monitoring posts on the plant's premises, producing isolated (and hard-to-interpret) numerical figures that were continuously made available to the public.[9] As events unfolded over the following days, such anxieties would only grow stronger.

Around March 15, as the buildings housing reactors 2 and 4 exploded within the span of a single day, mistrust in the Japanese government's style of information (non)disclosure started to gain full momentum. One important case in point was the suspected (and subsequently confirmed) government mishandling of its System for Prediction of Environmental Emergency Dose Information (SPEEDI), a simulation system under the jurisdiction of the Ministry of Education, Culture, Sports, Science and Technology (MEXT) to forecast radiation dose diffusion rates during a nuclear accident. The system had started simulating the direction and range of radioactive emissions on March 11, and immediately submitted a number of technical reports to the Nuclear and Industrial Safety Agency (NISA) and other relevant agencies. However, as became clear on March 23 when they were finally released, those reports had somehow been lost in the bureaucratic labyrinth before reaching cabinet members (Fukushima Genpatsu Jiko Dokuritsu Kensho Iinkai 2012). Because the public knew of the existence of SPEEDI, its conspicuous silence in the intervening days, between March 15 and March 23, served to create strong public impressions of the government's information manipulation.

A wider public sense of bureaucratic incapacity was also spurred on by the badly coordinated information disclosure practices across government and industry agencies, including the fact that press conferences were held separately by NISA and TEPCO. Besides being full of esoteric and technical expert jargon, press releases from these two main agencies occasionally contradicted each other, leaving many people searching for alternative sources of information. Moreover, the specific material circumstances of the nuclear disaster created a situation ripe for inference, speculation, and rumor. With no human access to the reactor buildings possible, and devices set up to indicate their safety status rendered unreliable by successive rounds of damage, the authorities continuously failed to establish a coherent frame that could contain the many ongoing symbolic (rumors) and material (radioactive) 'overflows' (cf. Callon 1998).

In the media space, such multiple uncertainties incarnated as a discrepancy between domestic and foreign media reports. In contrast to the Japanese media and their almost total dependency on government press releases, foreign media was trying to access firsthand sources of evidence. This method seemed vindicated when the British Broadcasting Corporation (BBC) managed to capture on video the critical moment of the reactor building explosion on March 12. These images—fed back to the Japanese public via digital media—proved a shock to many Japanese citizens. Long-standing institutionalized practices of how the Japanese state shores up its own technoscientific authority, tied to government control of media-borne information flows, was starting to come undone in the face of pressures from foreign journalists, new digital media, and the unruly behavior of an out-of-control 'decomposing' nuclear power plant.

In the following days, the disjunction between versions of reality conveyed by domestic and foreign media only became wider. At the same time, speculations about a 'hidden reality' behind the possibly fatal situation at Fukushima Daiichi multiplied. Widely divergent stories about the disaster started circulating on the Japanese-language Internet, exerting a persuasive force on people (Ogiue 2011). Public suspicion was boosted by a growing alienation from the uniform style of Japanese mainstream media reportage, which merely repeated official press releases of still more formulaic-sounding government announcements that things were 'currently under investigation'. Given widespread suspicions that TEPCO and the government were hiding 'unfavorable' information underneath such formulas, and since mainstream Japanese journalists did little to investigate their claims, trust in domestic television and newspaper sources gradually became widely undermined.

In this situation, concerned Japanese citizens turned to Internet media and social networking services (SNS) more than ever before to communicate about just what was happening. The increasingly widespread reliance on digital media was thus, at least partly, in response to growing disillusionment with mainstream media and their too-close reliance on official government sources, through what is known as the *kisha kurabu* ('reporter's club') institution (Takeshita and Ida 2009). On this point, the Fukushima Daiichi disaster reactivated ongoing tensions in Japan. In the months prior to the earthquake, freelance journalists had been criticizing the exclusive press clubs of central ministries, attracting attention among Internet and SNS users. During the disaster, such outspoken critics and their growing digital constituencies found more evidence of the tacit collusion between industry, bureaucracy, and mainstream media, as press clubs would habitually exclude freelance and foreign journalists who might challenge the 'officially' negotiated version of nuclear risk realities (see also Hara in this book).

To Japanese citizens, uncertainties were further compounded by the tragic singularity of the combined earthquake, tsunami, and nuclear disaster, spurring a widespread sense of disrupted realities. In this dramatic juxtaposition of extraordinary events, public attention was constantly torn, and public concerns multiplied chaotically. Only gradually did some of these concerns come to be anchored in shared practices of civic engagement, helping to restore a grip on an otherwise fractured reality. It is one of these emergent civic coordination efforts that we turn to now.

EMERGENT CONCERNED GROUPS: THE CIVIC RADIATION MONITORING MAP

Faced with different versions of a potentially fatal situation, a group of concerned citizens started sharing radiation monitoring information on the web. The person initiating this move was a video director, with a Tokyo-based career behind him, who currently runs a video production company in Nagano. On March 13, at a very early stage in the disaster, he used his Twitter account, named 'MFkurochan', to call for people's collaboration in building a Google Map covering radiation monitoring posts all over the country.[10] As he started creating the map, it soon attracted more followers, over time becoming one of the most important tools for citizens concerned by radiation risks to take stock of the situation.[11] The map came to serve as the infrastructural node of an emerging sociotechnical network, formed through the ad hoc collaboration of 'amateur' citizens with no formal scientific credentials but with practical knowledge gained through collective learning processes.[12]

As sociologist Ulrich Beck (1992) noted in his well-known work on 'risk society', unlike industrial air pollution, symbolic of a bygone era in present-day Europe (and indeed Japan), radiation epitomizes the 'deprivation of the human senses' in the age of nuclear catastrophes. Confronted with nuclear radiation, one is completely dependent on some form of technoscientific mediation; the threats, while possibly very real, simply cannot be touched, seen, or smelled by the human sensory apparatus. This fundamental fact is important, because it underlies the deep sense of fractured realities, and the sense of alienation from one's own bodily engagement with the world, that emerged within Japan in the wake of the Fukushima Daiichi disaster. For many concerned Japanese citizens, the mundane action of walking outside in the rain turned into a reflection on long-term fatalities: Could this affect my body somehow? Perhaps give me cancer in 25 years?

In this context of spreading anxieties, the emerging infrastructure of a civic radiation monitoring map would gradually come to assist people in structuring what to 'sense' in their immediate surroundings. On his Twitter timeline, the Nagano video director indicated that, as part of his inspiration for the map, he consulted a manual on civil defense and disaster mitigation called *Civil Defense*, published by the Federal Office of Civil Protection and Disaster Assistance in Switzerland. This anecdotal connection is telling. Less than impressed by his own government's handling of the nuclear disaster situation, the Nagano video director turned to guidelines assembled in a faraway country, inscribed with quite different disaster scenarios, but embedded within an overarching commitment to protecting and assisting the civic population. Rather than entrust this work to the Japanese equivalents of Swiss disaster authorities, given the sense of government incapacities during the nuclear disaster, he decided to rely on civic mobilization.

Via the MFkurochan Twitter account, calls were issued for people to collaborate in the construction of the civic map, and for volunteers to monitor radiation levels in their own vicinities. Several people responded to this request and started reporting on the Internet the counts per minute (CPM) values of Geiger counters, to which they happened to have access (cf. Kera et al. in this book). The map then came to function as a portal site for such civic monitoring posts. Often, the identities of those doing the actual monitoring were not entirely clear. One notable example, however, was a British biologist, teaching at the International Christian University in Tokyo, who had access to the Geiger counters in his laboratory and started measuring daily radiation levels on the campus shortly after the incident, releasing this data on his personal website.

At the same time, the map also came to reveal that there had already been a number of public and private monitoring posts put in place across the country long before the disaster. There were monitoring posts set up and run by prefectural governments all across Japan, coinciding with the locations of the country's more than 50 nuclear power plants. At these sites, prefectural governments established radiation research and monitoring centers, to oversee the impact of nuclear-related facilities on the environment, and to make data on air radiation dose rates available to the public. Websites of these centers often take the shape of a local area map, showing the location of monitoring posts, the positional relation between posts and plants, as well as wind direction and weather information. Similarly, TEPCO and other power companies had for a while been releasing their radiation monitoring data from plant premises on their websites. However, in spite of the rich information available on the prefectural monitoring websites, few in the general public knew of their existence, nor were the sites related to each other, before the civic map construction effort started.

In addition to these public and industry monitoring posts, the map revealed a few private monitoring posts already set up before the incident. The most important of these was a meteorological observatory run by Ishikawa Hiroshi, a senior IT engineer. His was one of only a few long-time private monitoring posts providing well-calibrated data, together with detailed information on meteorological conditions which might affect radiation levels. The monitoring history of Ishikawa goes back to the year 2000, when he constructed an eco-house in Hino city in the western suburbs of Tokyo, which utilizes solar light and natural wind for air conditioning. This prompted his interest in measuring the natural energy utilized by the house, which gradually led him to establish a small meteorological observatory. He added radiation monitoring to his observatory in 2005, when tensions over North Korea's nuclear tests were running high. Following the Fukushima disaster, his website attracted huge attention, with more than 60,000 hits on March 15, 2011 (Ishikawa 2012).

Over time, the civic map came to shine light on not only the new ad hoc measurement efforts of citizens, but also a range of already existing radiation monitoring posts run by municipalities, companies, research institutes, and individuals. On March 21, for instance, the map consisted of links to 64 monitoring posts, with 23 run by provincial or municipal governments, 15 by utility companies and related organizations, 10 by the central government, eight by independent citizens (including 1 private company), and eight by universities and research institutes (as indicated in Figure 5.1). As the civic radiation monitoring map expanded in scope, it also attracted

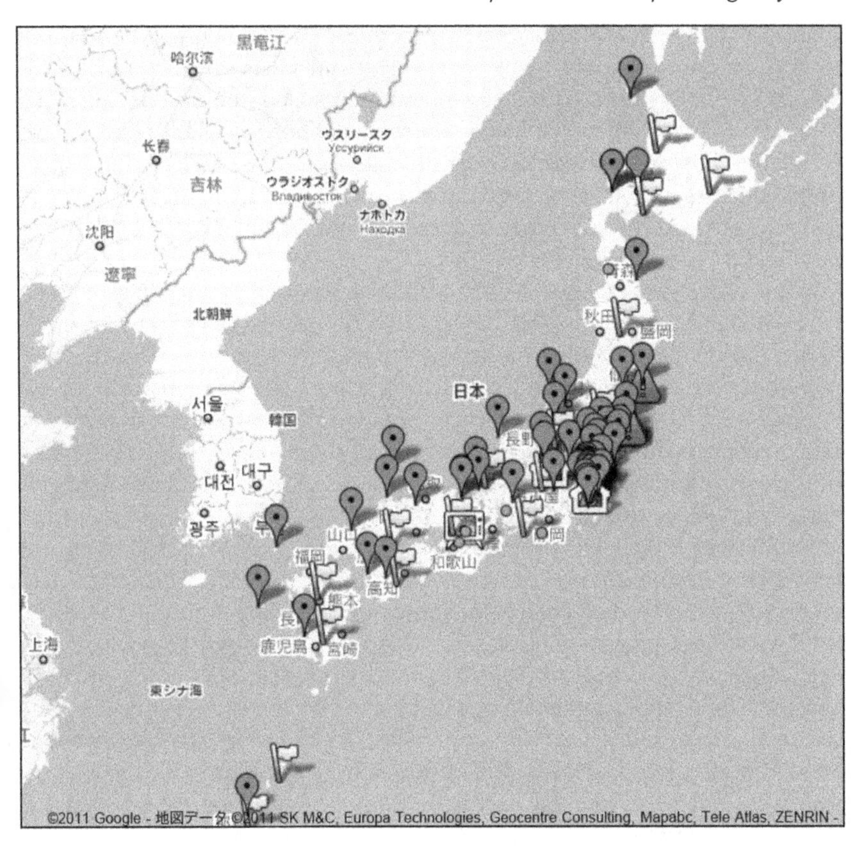

Figure 5.1 The civic radiation monitoring map captured on March 25, 2011

increasing and vast numbers of users: by March 21, the map had received a total of 3.9 million hits, and four days later the traffic surpassed 5 million.[13]

Amidst emergency conditions, the civic map thus sought to plot the full variety of radiation monitoring posts, as run by diverse organizations, and which were accessible via the web. In this sense, the map came to function as a link between heterogeneous data infrastructures, thereby 'drawing things together' in a single radiation inscription (Latour 1990). Also, by establishing new linkages among heterogeneous social identities, like other emergent concerned groups (cf. Callon and Rabeharisoa 2008), participants in this ad hoc civic radiation monitoring network actively sought to blur distinctions between 'private' and 'public' spheres.

Heterogeneity, in this context, meant more than simply data expansion; it also played a crucial part in establishing the credibility of the map as such. Hence, given the overall mistrust in governmental information disclosure practices, existing posts run by power companies and local governments often became targets of suspicion. However, as the civic map was able to juxtapose all the monitoring posts, regardless of their ownership, it was possible to compare data from government-run posts with those of individuals or universities. Within this juxtaposed space, the monitoring done by amateur citizens played a significant role to counter credibility problems, as reliable sources for crosschecking the data of government- and industry-affiliated posts. As an emergent infrastructural effect, the civic map thus achieved qualities of coherence and reliability that none of its constituent parts could possess individually.

In fact, engaged citizens seemed well aware of the predicament of mistrust in the existing monitoring posts, and actively tried to build up credibility around their own radiation measurements. They did this, for instance, by making the conditions under which measurements were made as transparent for others as possible. Hence, they would provide information on such 'details' as the technical specifications of their Geiger counter; its exact location; proximity to building structures; and other issues that might potentially influence the data. As such, one might say that the amateurs tried to counterweigh their lack of certified expertise by way of efforts to ensure transparency around their 'experimental' setups. In effect, this strict civic 'protocol' also served to create the impression that direct access was being provided to raw, unmanipulated data. This was particularly true where amateur citizens would upload their live, hour-per-hour Geiger counts directly on USTREAM, the Internet broadcasting site. Within a week, such broadcasts had a huge following, with one USTREAM account of a Geiger counter in Ota city, Southwestern Tokyo, receiving more than 630,000 hits by March 20. At 2 p.m. that day, nearly 2,000 people watched the broadcast.[14]

At the same time, new developments were already underway. Within just days of the civic map's initial establishment, some of the engaged citizens, including those broadcasting their counts on USTREAM, gradually ceased their monitoring activities again. This happened not as a sign of 'civic fatigue', but in explicit acknowledgment that other, better-equipped entities would now take over the task. On March 18, for instance, one USTREAM broadcaster in Koto city in eastern Tokyo announced the termination of his post, referencing that MEXT, together with various universities, had now established access to public monitoring posts, thereby taking over as independent trustworthy monitors.[15] In this sense, having established the civic map as an infrastructural node, citizens then performed an ad hoc act of public delegation, designating universities (and MEXT) as reliable spokespersons of this sociotechnical network. Just as Callon and Rabeharisoa (2008) show in a different context, this emergent concerned group was thus active in reconfiguring how political delegation worked in the post-Fukushima emergency situation.

For instance, in a new move, the University of Tokyo and Tohoku University launched their own monitoring projects, based on their campuses, within a week of the disaster and started releasing the data publicly. This data was later integrated into the civic radiation monitoring map. Tohoku University, located in the city of Sendai just 100 kilometers away from the Fukushima Daiichi plant, started releasing monitoring data on March 18—the same day the USTREAM broadcaster ended his post.[16] The University of Tokyo started releasing monitoring data, produced from its three campuses, via its internal website on March 15, and soon transferred this to its public website.[17] Apart from providing transparency as to measurement conditions, this website also provided information on the possible health impacts of radiation, including acknowledging expert controversy around 'acceptable' levels of human exposure.[18]

Citizen participants in the monitoring network would widely regard such new university-driven posts as sources of reliable information. At the same time, as individuals now had more means of checking the internal coherences among a wider range of heterogeneous monitoring data, trust could gradually be reestablished also in the monitoring posts of municipalities and power companies. This restoration of trust in the more 'official' radiation monitoring infrastructure, led to a gradual decline in the number of citizen-driven posts. However, active citizen-driven posts, installed before or after the disaster, still existed at the time of writing. The role of such citizen-driven posts also gradually shifted, as needs arose for more 'purpose-tailored' information, such as from around specific localities (e.g. a nursery) or for tracking food-related radiation issues (see also Kera et al. in this book).

In short, as a constantly evolving civic-technological object, the map initially built by MFkurochan would absorb this changing institutional landscape of monitoring posts, and provide hyperlinks to websites as new contributors joined the network. Although the monitoring posts covered by the civic map were initially mainly focused on the Tohoku and Kanto areas, it soon expanded to include most parts of the country. As such, the map became an instantly available and effective means for citizens in diverse geographical locations to gain access to information about radiation threats, and to take personal precautions as they saw fit. The heterogeneity of monitoring posts, as run by divergent entities with often contradictory interests, rendered the civic map into an 'objective' and credible tool for navigating an otherwise highly uncertain and 'risky' landscape. In the indeterminate state of emergency that shook the trust of existing data sources, credibility was thus slowly rebuilt from the ground up. This was done through the contribution of an emerging and constructive civic environmental infrastructure *appropriate* to the immediate material, political, and technological realities of post-Fukushima Japan (cf. Fortun 2004).

Moreover, the map also became an important tool for citizens to evaluate the plausibility of both official government statements seemingly aiming to reassure, and the many free-flowing rumors on pending catastrophes. In the general mood of mistrust during the initial stages of the disaster, it was particularly important for people to be able to validate the officially announced radiation levels, by checking them against the citizen-led data infrastructure. Just as importantly, the civic map also provided a practical means of assessing, in real time, the many divergent 'unofficial' stories circulating via Internet, Twitter, and foreign media. Most of these unofficial stories would conjure a 'hidden reality' of undisclosed events at the plant, suggesting the potential for catastrophic events to happen in densely populated cities such as Sendai and Tokyo. However, with effective monitoring in place, such catastrophic events would seem more easily detectable to concerned citizens. As such, the civic radiation monitoring map was instrumental in restoring some public sense of a *grip* on an otherwise fractured, multiplying, and anxiety-generating reality.

INFRASTRUCTURAL INVERSION: MONITORING MADE PUBLIC

One important theme permeating STS work on the social and organizational dimensions of technical infrastructure is that such infrastructures tend to be taken for granted, forgotten, and remain invisible (Star 1999). Nuclear radiation monitoring systems in Japan provide a clear illustration: prior to the Fukushima Daiichi disaster, this technoscientific information infrastructure was entirely in the background and unknown to most Japanese citizens. Atmospheric radiation is not a common object of public concern, and data on its fluctuations would be hard to come by for most citizens. Only seldom is the fact of radiation elevated to public attention.

Nevertheless, as Susan Leigh Star (1999) pointed out, the invisibility of infrastructure is never complete; rather, there is a *relational* quality to any infrastructure's hidden existence. One person's infrastructure is another person's focus of attention and work: for the cook, the water flowing in the kitchen sink is a taken-for-granted part of making a dinner; for the plumber, however, it is a target of constant repair and maintenance practices. This relational quality of infrastructure is important to understand the role of the civic monitoring map in gradually *reestablishing* citizen trust in the existing network of government-sponsored radiation monitoring posts. In a dramatic sense of the term repair, the work undertaken by the emergent concerned group of civic radiation monitoring may be understood as 'repairing' one small aspect of an otherwise shattered reality of multiple failing infrastructures. In the language of Callon (1998), the civic radiation monitoring map was thus part of an effort to *reframe* the emergency, by taking into account public concerns with radiation overflows.

To allow this work of civic infrastructural repair, in terms of enhancing the reliability of available radiation information, we suggest that the crucial aspect of the civic monitoring map was the way it helped citizens undertake what Bowker (1994) calls an 'infrastructural inversion'. In general terms, Bowker and Star (1999: 34) define infrastructural inversion as a procedure whereby people—including STS scholars—learn 'to look closely at technologies and arrangements that, by design and by habit, tend to fade into the woodwork'. 'Looking closely' at the technologies and arrangements of radiation monitoring, as a hither-to back-grounded reality, is exactly what the civic map allowed its co-emerging group of concerned citizens. In making the inner workings of this infrastructure visible, the map essentially worked as a tool of civic engagement, allowing its users to quickly learn enough about nuclear radiation monitoring to intervene sensibly (and 'appropriately') in the situation.

Clearly, part of the infrastructural inversion was prompted by the Fukushima Daiichi disaster itself, in directing people's attention to the immediate importance of nuclear radiation monitoring. As is well known in the STS field, back-grounded infrastructures tend to only become visible when they break down (Star 1999). However, this is only half of the story. For the inversion to happen, and for civic repair work to set in, citizens needed to *actively* make the failing infrastructure a topic of investigation, evaluation and reworking. In this context, lack of trust in the government's information (non)disclosure played an important role, in that it also spurred on an active mistrust in the existing monitoring infrastructure, setting in train a collective search for more reliable and more actionable information. It was the acutely experienced *necessity* of knowing the mechanisms of radiation measurement, to gain some grip on divergent realities, which drove citizens to undertake the infrastructural inversion involved in setting up a civic radiation monitoring map.

In this context, the monitoring posts installed individually by citizens played a crucial role, as a kind of 'experimental' device allowing citizens to learn about the inner workings of radiation monitoring equipment. Given the overall mood of distrust, citizens tried to attain transparency and 'objectivity' in measurements by disclosing the procedures and conditions of data generation as fully as possible. As a result, the citizen-run monitoring posts served to collectively unpack a technical 'black box' of radiation monitoring (Latour 1990). Among people actively involved, efforts to make their own measurements transparent could then 'reveal', by way of extrapolation, some of the internal workings of existing government- and industry-run posts. To engaged citizens, a handful of individual posts thus came to serve as experimental 'exemplars' for the vast network of innumerable, and inaccessible, public monitoring posts. As Star (1999) notes, most infrastructure is big, layered, and complex; repair of infrastructure is always partial, never total. This indeed seems an important lesson to learn from the emergent concerned group of radiation monitoring in post-Fukushima Japan.

In the process of drawing heterogeneous data infrastructures together into the single space of what became the civic map, citizens gradually came—within the confined microcosmos of radiation monitoring—to rearrange the material, technical, and political conditions under which the crisis of credibility in post-Fukushima Japan could be slowly and partially reworked. Part of this process, as noted, involved an active process of juxtaposing and calibrating data, as the measurements coming from citizen-led and government-sponsored posts were checked for their internal coherence. In this context, the achievement of coherence would signal an important *absence* of official manipulation, thereby gradually also allowing for a partial reestablishment of trust in the more 'official' parts of the monitoring infrastructure.

Perhaps we should speak here of an infrastructural 'reversion': in inversing the existing radiation monitoring network, experimenting on it, adding new elements to it, and testing it for official manipulation, citizens helped to actively 'reverse' the now reconstituted infrastructure to the level of practical workability for individual people. While public criticisms of government and TEPCO policies would obviously continue (on any number of grounds, radiation-related and otherwise), the 'reversed' radiation monitoring infrastructure served to reestablish a minimal amount of 'shared reality' in an otherwise quite chaotic situation.

In short, far from being the sole prerogative of technical experts, government officials, electricity companies, or, indeed, STS scholars, our analysis suggests that infrastructural inversions—and the crucial learning, infrastructure repair work, and collective restructuring they facilitate—may also well fall within the capacities of informed, engaged, and concerned citizens acting together in engaging more official institutions of power. Indeed, and more strongly, particular situations of emergency may call for such civic capacities, given the way established institutional frameworks fail, collapse, or otherwise prove inadequate in addressing collectively experienced infrastructural problems. While humble acts of installing radiation monitoring posts and plotting data onto a digital map may have no place in grand narratives of catastrophe and doom, we suggest they are nonetheless critical as a lens into the robust public intelligence at work all over post-Fukushima Japan.

CONCLUSION: COLLECTIVE LEARNING THROUGH EMERGENCY?

In this chapter, we address one of the many 'minor' histories that we believe should be told, amidst wider ongoing social, political, and technological reflections, on the world-historical events of the Fukushima Daiichi disaster. Our focus on the practical activities of an emergent group of Japanese citizens, who deploy technologies to address their collective sense of concern with atmospheric radiation risks, highlights the importance of mundane situated ways of knowing and acting together in the face of emergency. As an ad hoc public brought together by the contingencies of the situation, we suggest that civic collaboration on radiation measurement became one project amongst a variety of citizen-driven projects to gradually fill in the serious void of technoscientific credibility that Fukushima Daiichi left open within Japan. More importantly, we claim, the infrastructure assembled in a civic radiation monitoring map proved an appropriate tool, amidst a fractured sense of multiplying threats, for citizens to regain some grip on collective reality.

We situate these claims within wider discussions—well-known, albeit to different effects, throughout the whole of STS studies—on the critical importance of civic engagement to any well-functioning technoscientific democracy (Brown 2009, also Hindmarsh in this book: chapter 4). More specifically, we suggest that the pragmatist lineage of political philosophy, as embedded deeply in some recent ANT approaches (e.g. Callon and Rabeharisoa 2008; Latour 2007; Marres 2007), is particularly well suited for giving ontological and political weight to the work of collective creativity, intelligence, and restructuring undertaken by emergent publics in the wake of the Fukushima Daiichi disaster. As a supplement to deliberative approaches, which tend to inscribe civic engagement into new arrangements of representative democracy, pragmatism has the virtue of stressing that public engagement with technoscience is called for most strongly at exactly those moments when established political and institutional frameworks *fail* to address collective concerns.

The state of emergency experienced in Japan in the immediate post-Fukushima days—characterized by extreme levels of technoscientific uncertainty, controversy, disorientation, and mistrust in government—in many ways epitomizes this pragmatist sense of a highly problematic public situation. In the language of Callon (1998), nuclear emergencies are situations of pure 'overflow'. These overflows are part material, as radioactive emissions spread beyond any control, and part semiotic, as people's taken-for-granted senses of reality threaten to unravel to the point of existential disruption. In this context, we claim, the emergent concerned group of radiation monitoring amateurs managed to steer a narrow path between two equally unattractive alternatives: on the one hand, the formulaic and manipulative power of continuing government reassurances; and, on the other, the dense webs of catastrophe rumors circulating through digital and foreign media. The insistence on continual observation, cautious measurements and public information-sharing as demonstrated by these amateurs, we believe, manifests a robust sense of collective intelligence under difficult circumstances.

In this context, it is interesting—if also ultimately very difficult—to speculate further on the contextual historical, political, technological, and cultural ramifications of this, and other, Japanese public engagements with technoscience in the wake of the disaster at Fukushima Daiichi. Throughout this chapter, we have hinted in this direction via the category of civic epistemology (as civic knowledge practices), borrowed from Jasanoff (2005) to suggest some of the characteristic ways in which reactions to the Fukushima Daiichi disaster both reflected, and seriously challenged, long-standing (and rather technocratic) practices for shoring up the public credibility of technoscience in Japan (cf. Kimura 2010).[19] In particular, we have stressed how the strong public reliance on digital media formats, expressed at the core of the emerging civic radiation monitoring map, represented a widespread sense of critique and disillusionment with standard government practices of information (non)disclosure, especially through the exercise of control over mainstream media. It seems reasonable to suggest that such critical tension, and the governmental credibility 'deficits' it signals, will continue, and likely extend into other domains of Japanese technoscientific politics.

On a somewhat different note, what does our case study on public engagement imply or suggest about the key issue of the possible future(s) of nuclear power in Japan? Methodically speaking, the most sober answer would seem to be: probably not all that much—although, at some level, the emergent concerned group of radiation monitoring amateurs may be taken to represent one amongst a plurality of mushrooming nuclear-anxious publics in contemporary Japan. However, the extent to which this widely registered sense of newfound public unease, manifested also in opinion polls, will translate into active antinuclear opposition is a somewhat different issue. As a matter of informed judgment, we believe that, just as responses at the level of official government politics are still very much unsettled, Japanese public opinion on nuclear power is currently in an ambiguous state, caught in between a widespread sense of passive unease, a mostly silent desire for nuclear phaseout, and smaller pockets of more outspoken antinuclear rejection. It seems that much still depends on the way new issues, revelations, and controversies related to the Fukushima Daiichi disaster will keep mobilizing public concerns for some time yet into the future.

However, what does seem certain from our discussion is that considerable potential and capacities exist for technoscientific creativity and informed collective learning processes in the Japanese public, sensitized to the threats of nuclear disaster. In their attempts to reassemble the political, economic, material and cultural forces of the country in the wake of disaster, Japanese technoscientific and other elites would seem well advised to tap into these public resources. Important as this certainly is, the issue here is, however, not primarily whether Japanese bureaucratic authorities will manage to break with their own somewhat technocratic legacies, by instituting more open-ended forms of public dialogue and consultation on nuclear and other technoscientific futures. More to the pragmatist point, the issue is whether the many proliferating publics, spurred into being by concerns for a safer, greener, more equitable and sustainable technological future, will be allowed to deploy their problem-solving capacities for the greater good of the Japanese 'risk society' (cf. Beck 1992). In this respect, we suggest, we could do worse than adopt the emerging concerned group of radiation monitoring amateurs as an important component of a blueprint for change.

NOTES

1. For an excellent introduction to ANT, see Mol (2010). Marres (2007) provides an elaborate discussion of how ANT relates to pragmatist political philosophy.

2. Throughout the chapter, 'risk' is taken not in a technical-rational sense of 'controlling chance', but rather in the sociological sense of shared uncertainty, non-knowledge, and anticipation of future catastrophes as implied in Beck (1992).
3. By 'virtual' or 'online' ethnography, we refer to the use of participant-observation to explore communication and networking in the Internet space (see Hine 2000).
4. In accordance with standard usage, 'auto-ethnography' denotes how we, as researchers, attempt to link our personal experiences (of the disaster) to wider cultural and political frames of meaning.
5. Needless to say, there are and have been widely different reactions to the Fukushima disaster in Japan, reflecting varying senses of urgency and danger. On this point, the fact that we were personally located in the western parts of Japan makes a real difference. In future work, it will be important to situate Fukushima in its wider Japanese political-historical context. While this is beyond the scope of our chapter, see Yoshioka (2011) and Kainuma (2011).
6. In Japanese politics, Prime Minister Noda officially declared an end to the Fukushima disaster on December 16, 2011. However, in the wider sense of the technopolitics of nuclear power, the event clearly remains unsettled.
7. For a fuller discussion of this notion of 'matter-of-concern', and how it relates to a pragmatist-inspired notion of democracy, see Latour (2005).
8. In fact, the Fukushima governor announced a two-kilometer evacuation directive, thus marking the first of a long series of internally contradictory announcements from the authorities.
9. As we explain later in the text, the civic activities of radiation monitoring did not create their observation network from scratch, but rather sought to coordinate existing government-sponsored and newly created civic-run monitoring posts into a more comprehensive picture.
10. In the course of events, several experts—including atomic physicist Ryogo Hayano and radiologist Keiichi Nakagawa—transmitted technical knowledge via their Twitter accounts. As a 'layperson', however, MFkurochan did not aim to transmit his own knowledge but instead used his account to build collaboration on the civic radiation monitoring map.
11. Readers may still be able to access the civic radiation monitoring map at: http://maps.google.co.jp/maps/ms?ie=UTF8&oe=UTF8&msa=0&msid=208563616382231148377.00049e573a435697c55e5, accessed May 24, 2012.
12. 'Amateur' obviously carries no derogatory connotations here but simply points to the fact that the engaged citizens were not 'professional' (i.e. trained) radiation monitoring experts.
13. All figures here stem from our (the authors') screen captures, taken on March 21 and March 25, 2011.
14. Figures from our (the authors') screen capture, March 20, 2011.
15. Announcement available at: http://www.ustream.tv/channel/geiger-counter-tokyo, accessed June 23, 2011.
16. Tohoku University's website, accessed March 7, 2012. http://www.bureau.tohoku.ac.jp/anzen/monitoring/english.html.
17. University of Tokyo website, accessed March 7, 2012. http://www.u-tokyo.ac.jp/ja/administration/erc/index.html.
18. For details on this particular controversy, see the University of Tokyo website at note 18.
19. We use the term 'technocratic' in a mostly descriptive sense here, to point to the heavy reliance on technical expertise within Japanese bureaucracies as such (see Kimura 2010).

REFERENCES

Beck, U. *Risk Society: Towards a New Modernity*. London: Sage, 1992.

Bowker, G. "Information Mythology and Infrastructure." In *Information Acumen: The Understanding and Use of Knowledge in Modern Business,* edited by L. Bud-Frierman, 231–247. London: Routledge, 1994.

Bowker, G., and S. Star. *Sorting Things Out: Classification and Its Consequences*. Cambridge, MA: MIT Press, 1999.

Brown, M. *Science in Democracy: Expertise, Institutions, and Representation*. Cambridge, MA: MIT Press, 2009.

Callon, M. "An Essay on Framing and Overflowing: Economic Externalities Revisited by Sociology." In *The Laws of the Markets,* edited by M. Callon, 244–269. Oxford: Blackwell Publishing, 1998.

Callon, M., and V. Rabeharisoa. "The Growing Engagement of Emergent Concerned Groups in Political and Economic Life: Lessons from the French Association of Neuromuscular Disease Patients." *Science, Technology & Human Values* 33, no. 2 (2008): 230–261.

Daston, L., and P. Galison. *Objectivity*. Cambridge, MA: MIT Press, 2007.

Fortun, K. "Environmental Information Systems as Appropriate Technology." *Design Issues* 20, no. 3 (2004): 54–65.

Fukushima Genpatsu Jiko Dokuritsu Kensho Iinkai (Independent Investigation Commission on the Fukushima Daiichi Nuclear Accident). *Chosa-Kensho Hokokusho (Investigation Report)*. Tokyo: Discovery 21, 2012.

Hine, C. *Virtual Ethnography*. London: Sage, 2000.

Ishikawa, H. "Saigai to Joho Koukai (Disaster and Information Disclosure)." *Journal of the ITU Association of Japan* 42, no. 1 (2012): 48–52.

Jasanoff, S. *Designs on Nature: Science and Democracy in Europe and the United States*. Princeton: Princeton University Press, 2005.

Kainuma, H. *"Fukushima" Ron: Genshiryokumura wa Naze Umareta no Ka (On Fukushima: Why the Nuclear Power Village was Born)*. Tokyo: Seitosha, 2011.

Kimura, Ayo Hirata. "Between Technocracy and Democracy: An Experimental Approach to Certification of Food Products by Japanese Consumer Cooperative Women." *Journal of Rural Studies* 26, no. 2 (2010): 130–140.

Latour, B. "Drawing Things Together." In *Representation in Scientific Practice,* edited by M. Lynch and S. Woolgar, 19–68. Cambridge, MA: MIT Press, 1990.

Latour, B. "Turning around Politics: A Note on Gerard de Vries' Paper." *Social Studies of Science* 37, no. 5 (2007): 811–820.

Marres, N. "The Issues Deserve More Credit: Pragmatist Contributions to the Study of Public Involvement in Controversy." *Social Studies of Science* 37, no. 5 (2007): 759–780.

Mol, A. "Actor-Network Theory: Sensitive Terms and Enduring Tensions." *Kölner Zeitschrift für Soziologie und Sozialpsychologie* 50, no. 1 (2010): 253–269.

Ogiue, C. *Kensho Higashi Nihon Daishinsai no Ryugen, Dema (Examining Rumors on East Japan Great Earthquake)*. Tokyo: Kobunsha, 2011.

Star, S. "The Ethnography of Infrastructure." *American Behavioral Scientist* 43, no. 3 (1999): 377–391.

Takeshita, T., and M. Ida. "Political Communication in Japan." In *Political Communication in Asia,* edited by L. Willnat and A. Aw, 154–175. London: Routledge, 2009.

Yoshioka, H. 2011. *Genshiryoku no Shakaishi (Social history of Nuclear Power)*. 2nd ed. Tokyo: Asahi Shimbun Publications, 2011.

6 Post-apocalyptic Citizenship and Humanitarian Hardware

Denisa Kera, Jan Rod, and Radka Peterova

Collection and sharing of sensor data by citizens over social networking sites is commonly described as 'participatory sensing' (as discussed in detail below).[1] The reasons why citizens get involved and provide data about their environment can vary from curiosity, self-interest and notions of citizenship to support of ecological awareness. In the case of the Fukushima Daiichi disaster, the main focus for such reasons became crisis monitoring and response in relation to potential radiation pollution. Collective online mapping during crises is not a new phenomenon. Before Fukushima it was mainly deployed in developing countries (e.g. Sudan, Haiti and Thailand). Tools for online mapping, such as the Ushahidi platform,[2] or new organizations devoted to the improvement of new 'humanitarian technologies', such as Crisis Mappers: the Humanitarian Technology Network,[3] have proved very useful for humanitarian projects in countries with limited resources, nonexistent infrastructure and no or limited government emergency response teams.

But the Fukushima disaster set an important precedent in terms of 'participatory monitoring' of 'technological crises'. It stressed not only the importance of independent (radiation) data gathering but also new sociotechnical practices of citizens building and investing in their own tools, which made this participatory monitoring unique in terms of citizen mobilization over data. The collective practices of investing in and building do-it-yourself (DIY) tools for radiation monitoring demonstrated a whole new dimension of citizen empowerment, which goes beyond issues of data to introduce prototype building as a new type of collective and political action. In other words, DIY Geiger counters and other post-Fukushima humanitarian tools enabled a unique case of citizen mobilization immediately following a disaster. The independent collection and crowdsourcing of data together with building open source and DIY tools became a form of a 'political right', similar to the concepts of freedom of speech and information (Peterova 2011), which aim to strengthen individual but also collective understanding and responsibility, in this case, for the environment and human health.

In this context, this chapter reports on our study of three notable activities of participatory radiation monitoring concerning the Hackerspace movement and development of open source hardware prototypes in Japan in relation to the Fukushima disaster. The Hackerspace movement comprises alternative and independent research and development centers known as 'Hackerspaces', as physical spaces in a global network in which citizens work on independent software and hardware projects funded by themselves or by way of online (crowdfunding) tools and the leveraging of skills and resources. These creative environments support so-called 'hacking', which means modification and building of open source tools, and any form of appropriation of technologies, for freely defined purposes by citizens and users themselves rather than by any corporate or business entities.

In our study, we used web ethnographic data as well as survey and interviews with the participants of two projects, Safecast,[4] and Radiation Watch,[5] as conducted online between October and December 2011. We aimed to better understand the motivations and experiences of the participants in relation to the nuclear disaster as it unfolded. We find their actions inform a new type of citizenship—'cosmopolitical citizenship'—emerging around open source tools. We first followed the Safecast project, closely related to Tokyo Hackerspace, which has a global agenda of providing citizens with accurate and independent data on radiation. Secondly, we looked at Tokyo Hackerspace workshops on designing and building DIY Geiger counters and other radiation monitoring tools without Tokyo Hackerspace suggesting any explicit support platform for the data. This left it up to participants to decide which platform they preferred to share data on. Thirdly, we explored the popular Radiation Watch project, started by a few Japanese designers and engineers with cheap and simple DIY kits without expensive Geiger tubes. These tools were later developed into a product that anyone could buy and use with a mobile application for sharing data.

These three examples helped us identify the importance and the effects of these new civic sociotechnical practices in relation to the notion of cosmopolitical citizenship in an almost post-apocalyptic situation. The social networking sites, but also the DIY tools and open source hardware practices, that arose during the course of the Fukushima Daiichi disaster enabled citizens who accessed them to better deal with the crisis and the limited information broadcast by the official response team, on a very personal and day-to-day basis (cf. Morita et al. in this book). The social, political and technological aspects of this citizen mobilization employing DIY open source tools and online services for sharing data also stressed the notion of responsible and active citizenship. That is, responsible citizenship was performed through DIY practices of building tools and sharing data independently of both governmental and nongovernmental bodies.

In fields of knowledge study, this new grassroots response to a disaster brought together investigation of newly emergent social phenomenon, behavior and practices through science, technology and society (STS) studies on risk and uncertainty (e.g. Beck 1994, Giddens 1991), with science communication and policy studies on public participation in science and technology (e.g. Jasanoff 2003). The approach we decided to take to understand the convergence between citizen science projects and issues of uncertainty and risk after a disaster is mainly informed by actor network theory—or ANT (Law and Hassard 1999), as a useful framework for exploring the emergence and significance of DIY tools and open source hardware to enable this new type of citizen empowerment.

Actor network theory would view these novel sociotechnical tools as 'hybrid networks' involving human and nonhuman actors, which have equal weighting and ability to influence situational outcomes and processes. The outcome was that DIY hacking practices as forms of citizen mobilization and empowerment enabled more decentralized and novel forms of disaster management and emergency responses that aided citizens to better negotiate radiation pollution (cf. Schmid in this book). 'Citizenship' in such hybrid networks—rather than serving some predefined ideals of emancipation based on gender, social inclusion, or other socially defined concepts—is defined as an experiment in bringing together accessible technological possibilities with human interests and social needs.

The ANT concept of hybrid networks, based on a 'generalized principle of symmetry' between society, nature, and technology (Latour 1993), thus helps us better understand these emergent forms of public participation in science and technology, which we identify here as the Hackerspace and DIY movement. The need for independent data could not possibly be fulfilled without utilizing open source hardware tools. These tools were not originally conceived as social or political technologies, but as prototype devices for testing new design ideas, to be later manufactured on a larger scale. But the tools enabled an unprecedented form of citizen activism post-Fukushima. They entered and supported the everyday life and practices of a large number of lay users through workshops and online support groups changing traditional notions of citizen participation and involvement in crisis. The hybrid network created around Hackerspaces, which supported the building of DIY Geiger counters and sharing of independent data, gave equal agency and importance to the various actors in this post-apocalyptic situation.

The independent, citizen science activities postdisaster were political, social and technological interventions happening simultaneously. They depended on open source hardware tools as much as on active and motivated citizens. This is why we call them 'cosmopolitical' (Latour 2004) and refer to the more normative view of hybrid networks in ANT, which not only explains the events and the networks ex post but tries to devise questions on how to conjecture politics around new social actors. Building new tools as a way of introducing new actors in the case of this disaster enabled citizens to gather independent data on radiation directly and to create new networks and actions around this without governmental or nongovernmental help. That is, it was not a simple technical solution to a predefined and given problem but a normative 'event' and intervention, a process of empowering citizens to define new common goals and questions in a situation of crisis by testing these tools and prototypes.

The DIY activities around open source hardware thus enabled citizens to both challenge and build on the official radiation information being issued by government and nongovernment agencies post-Fukushima, which many commentators considered inadequate for understanding the spread and effects of radiation at the local, street and neighborhood levels (see also Morita et al. in this book). The resultant participatory radiation monitoring activities were not only a political venue of citizen protest and mobilization, but also served other functions, for example, as a type of ritual for gaining symbolic power over the circumstances, which improved people's morale and capacity to better face and cope with crisis. Building and using Geiger counters in some cases also reflected the act of creating and using a 'fetish' object (Kera 2012), which connected the concrete, material tool and radiation issue with deep seated fears, hopes, and emotions difficult to deal with. Overall, the various functions of these participatory DIY crowdsourcing practices extended the idea of community and responsible citizenship. The broader outcome was that post-Fukushima, traditional political and social characteristics of a responsible and caring community, such as voting, self-organization, freedom of expression, and solidarity, became extended through hybrid, sociotechnical practices, as illustrated in the building and application of radiation monitoring tools for data gathering and sharing of data.

OPEN SOURCE HARDWARE MOBILIZATION

The loose and hard-to-define citizens' network of geeks, citizens, and amateur scientists, which emerged post-Fukushima to address the threat of radiation pollution at the street level, showed for the first time the real potential of the so-called Hackerspace movement as a platform for grassroots politics and innovation.[6] Tokyo Hackerspace was especially quick to respond to the disaster to create a new model of crisis management. This model enabled concerned citizens to join forces with designers and members of the broader international Hackerspace community to build independent tools for gathering and sharing radiation data. Through a series of workshops and 'kits'—sets of simplified electronic parts, which can be assembled into a functional tool—Tokyo Hackerspace shared instructions with lay people on how to build and tinker with Geiger counters and radiation data.

This innovative case of participatory sensing of environmental data (e.g. of air pollution, soil pollution and water quality) extended the traditional notion of citizenship. In effect it started a 'citizen science' movement involving 'participatory monitoring of radiation'. Common participatory sensing and monitoring is an emergent trend in citizen science projects seeking greater involvement of lay people in participation and decision making about their habitat (Paulos et al. 2009). Alternatively, the postdisaster monitoring radiation network could be described as a grassroots and decentralized sociopolitical movement on radiation issues that emphasized cooperative data and knowledge sharing to enable small efforts to have larger impacts. When visualizing the collective results on a map—Google Maps (see also Morita et al. in this book)[7] or Google Earth[8]—or a visualization interface such as IBM Many Eyes,[9] a new sense of a community action and collective power was created.

But it could also be seen as a 'technological movement' in terms of its insistence on DIY open source tools distributed through kits,[10] and social software online services. In other words, the Fukushima grassroots radiation monitoring integrated crowdsourcing and participation over data with new technological trends—open source hardware and software platforms, such as Arduino,[11] or Pachube,[12] and other do-it-yourself (DIY) and do-it-with-others (DIWO) approaches for building DIY Geiger counters. These trends and approaches are part of Hackerspace culture and its connections to the *Make* magazine (from 2005), and Maker Fair events (from 2006),[13] which play an essential role in this emergent alternative research and action culture. The decentralized radiation monitoring by citizens over these DIY open source hardware tools changed the meaning of active and responsible citizenship from a communicative act (of discussing data) to a material practice (of building tools and prototypes). It served as a means of solving immediate and worrying problems related to lack of independent data and public knowledge about radiation concerning inadequate official information on radiation, to become a unique platform for citizen mobilization.

As such, we argue this new and emergent form of public participation on controversial technoscientific problems—here, on the spread and effects of nuclear radiation and what is the best course of action for citizens to address and cope with it—changes our views of what it means to be a citizen, which also helps to provide a resilient participatory approach for a possible post-apocalyptic world. In a world of the 'internet of things' (Gershenfeld et al. 2004, Greenfield 2006, Weiser 1993), and 'crowdsourced data', grassroots science citizenship appears to radically extend the notion of political participation based on communicative acts of discussing, deliberating and voting. It also involves 'giving voice' to nonhuman actors in our environment (Latour 1993, 2004), through sensors, open source hardware, and various online tools. More broadly, to reiterate, this sociotechnical citizenship can be described as 'cosmopolitical' (Latour 2004), as it defines public participation and civic engagement in terms of our involvement with nonhuman actors through building prototypes and using DIY open source tools to measure radiation risk. It connects politics and social interaction with building tools but also managing data from things and objects.

As such, post-Fukushima decision-making processes changed and transformed the elaborate divisions of expertise between government institutions, universities, research centers, and nonprofit organizations. At the local level, many citizens independently made small, individual decisions based on their Geiger counter measurements on how to move around, where and what to eat, and how to work and live under the constant threat and risk of radiation pollution. To best understand the emergence and significance of these micro decisions and policies of individuals and small communities we saw the need to discuss the public and private functions of DIY grassroots efforts and strategies related to participatory radiation monitoring. We therefore compared the available information and data available on the Internet about the two main initiatives and projects (participatory radiation sensing and Hackerspace workshops) with the interviews we conducted with citizens who took part in the radiation monitoring efforts of Safecast and Radiation Watch.

PARTICIPATORY SENSING OF RADIATION

The emergent practice of participatory sensing of radiation fits into the more established field of participatory environmental monitoring and sensing, as a standard citizen science practice with impact on both technoscience and policy (Hedgecock 2009, Paulos and Jenkins 2005). This field includes mobile sensing (Goldman et al. 2009), urban sensing (Campbell et al. 2006), and concepts of 'participatory urbanism' (Paulos and Jenkins 2005), which fits sensors into the design of cities and neighborhoods. Prior participatory sensing research has focused largely on air pollution (Hooker et al. 2007, Peterova 2011, Saavedra 2011), which seriously affects the quality of life in many cities. Hong Kong, Mexico City, Manila, and Beijing are outstanding contemporary examples.

It is relatively easy to 'sense' at the amateur DIY level because of low-cost sensors for CO^2, NO, NO^2, SO, and SO^2. The tools for participatory monitoring are usually 'explicit' (handheld devices) or 'passive' (vehicle-mounted sensors) and they record geo-referenced environmental data in places where people live or go, rather than where scientists might expect them to be (Hedgecock 2009). This usefully increases the amount of information on public life as it targets more specific areas that people feel personally connected with and in charge of. When neighborhoods start to independently monitor and collect data in real time, as was the case post-Fukushima, they also become more resilient in building their ability to handle the new hazardous atmosphere through more accessible tools to manage their immediate situation. Rather than waiting for the centralized authority to intervene and supply data, which was found in the Fukushima case to to take a long time and be inappropriate at the local level because of its technical complexity, citizens post-Fukushima acted and innovated at the grassroots level and demanded change based on user-friendly independent citizen data.

However, in the immediate postdisaster, there were neither enough Geiger counters nor other reliable tools to gather information on radiation levels in contaminated areas. Citizens had to start building their own devices and reflect upon their imperfections by calibrating them and figuring out their reliability. The resilient response of these citizens was apparent not only from crowdsourcing and the efforts to build tools, but also from financial support raised from the global community by way of crowdfunding websites.[14] 'Crowdfunding' is a relatively new term to describe public and global fundraising for innovative projects enabled through dedicated websites. Crowdfunded projects usually offer something in exchange, which, in the case of Fukushima, became an *actual* Geiger counter, used in workshops to teach people how to build and calibrate it to collect accurate and detailed data.[15]

Another problem addressed through independent radiation monitoring efforts was the inability of official radiation data to identify many random, high radiation concentrations called 'hotspots', which were sometimes quite hard to detect, for example:

> The hotspot, a small area of about one meter radius, was found in a vacant lot in Kashiwa. Radiation levels of 4.11 microsieverts per hour were detected one meter above the surface of the soil, equivalent to some areas in the evacuation zone around the crippled nuclear power plant. Up to 450,000 becquerels per kilogram of radioactive substances were detected in the soil below the surface, an Environment Ministry official said, Fuji TV reported. (Japan Today 2011)

Kashiwa is located 195 kilometers from Fukushima, on the outskirts of Tokyo in Chiba prefecture. Subsequently, hotspots became a key target of participatory radiation monitoring efforts. Official radiation data versions were also often late in providing information. Reliability was further undermined where fixed sensors, although accurate, were sparse, and the released data was often based on interpolation rather than measurement.

Weather conditions including sunlight, rain and wind facilitate atmospheric conditions for the formation of hotspots. While radiation travels with dust particles by wind or water, it tends to accumulate in drains and ditches both in natural and urban areas. Because of these unpredictable effects and hotspots, participatory radiation by citizens crossed the boundary of radiation monitoring being the sole domain of experts and policy makers. Citizens, in using their tools for this activity, effectively become 'ad hoc policy-makers themselves' in making decisions that would contribute to the health and safety of their streets and neighborhoods. The DIY Geiger counters thus became a type of 'technopolitical tool' (Goodman 2009), which transformed 'citizens as consumers of data' into active and responsible public actors providing highly localized information to their communities (see also Bäckstrand 2003).

HUMANITARIAN HACKERSPACE WORKSHOPS IN TOKYO

The transformation of consumers of data into public actors was done not only by data sharing practices but also through building and developing open source tools, as in the Tokyo Hackerspace workshops. The importance of the open source hardware as a novel response to the disaster was apparent from the online materials related to the Tokyo Hackerspace events (Akiba 2011, Reuseum 2011). Almost immediately after the Tohoku earthquake and tsunami, which informed Fukushima, and during the ensuing disasters, Hackerspace members held several meetings to discuss how to help affected areas. The first humanitarian hardware project of Tokyo Hackerspace, the Kimono Lantern Kit, was a solar rechargeable lamp originally designed as a decoration for gardens. It was quickly mass produced and distributed in areas suffering blackouts.

This form of local activism was later termed 'Humanitarian Open Source Hardware' (Akiba 2011). The lantern project served as a model for developing the subsequent DIY radiation sensing devices of various DIY Geiger counters (e.g. iGeigie, the iPhone Radiation Dock, and the Ion Chamber Radiation Detector Kit). The following is a description of one of the first Hackerspace meetings, which demonstrates the close connection between attempts to design open source hardware tools and activist ambitions to help affected areas. It expresses the motivation and empathy mixed with very practical concerns on what is needed in terms of technical, financial, and other support:

> In the hackerspace, we'll be holding our meeting tonight and will probably start hammering out plans to figure out how and where we can help. . . . So although it's outside the original sphere of intended use, it looks like the simple Kimono lanterns we designed can play a small role in providing comfort and at least give a small feeling of safety to people that are going through this horrific experience. . . . I've updated the files to v1.1 and the package includes the BOM and full gerbers. It's a turnkey package that can be taken and sent directly to the PCB fab. (Akiba 2011)

Bringing the open source lamp to the people in affected areas was thus not a simple utilitarian task but more an attempt to provide a feeling of safety after trauma, to give a feeling of personal control over the circumstances. It is an example of what is referred to as a 'fetish' function of humanitarian hardware, which is even more pronounced in the case of the DIY Geiger counters. Like indigenous culture fetishes, these objects have almost a 'magical power' to provide comfort in times of uncertainty and hazard. People monitoring radiation around their houses and neighborhood are aware that they are not protected from the physical effects of radiation but at least psychologically and mentally they feel protected from the uncertainty and chaos and hold onto the hope they can manage and improve their circumstances. The DIY radiation monitoring devices simply enabled a psychology of basic control and comfort (as well as practical outcomes), also related to a feeling that people were not alone but had the support of a global community.

The first Geiger counters were donated to Tokyo Hackerspace by Reuseum, a company with close connections to US Hackerspaces (Reuseum 2011).

After receiving the Geiger counters and improving their functionality, Tokyo Hackerspace distributed them to organic farming communities that needed accurate radiation data on their crops (Reuseum 2011). The Geiger counters were thus a global response of the Hackerspace movement and network, which demonstrated a belief system based on distributed and decentralized solutions that improved, in both a social and technical sense, the response offered by the Japanese government or the nongovernmental organizations concerning access to local radiation data (see also Morita et al. in this book).

The efforts organized by Tokyo Hackerspace were also connected to the RDTN.org network (later named Safecast[16]), which began its efforts a week after the nuclear meltdown. Safecast crowdsourced data over the Pachube platform to provide accurate sensor information to citizens about radiation levels issuing from the meltdown. Safecast also raised money through the crowdfunding website Kickstarter,[17] to buy more Geiger counters and create a sensor network providing finer-grained data as an addition to the official but again, unreliable stationary sensors. The data from the hacked Geiger counters and other DIY devices was then disseminated through a Google map and a Google earth layout and on various other platforms listed on the website Japan Geigermap.[18] These highly organized efforts thus emerged as a civic interest in how to best measure the immediate surroundings, identify radiation hotspots, and share and critically evaluate these measurements and their risk to health. The complex networks between Hackerspaces and these online services in the aftermath of Fukushima thus offers an important disaster case study showing the possibilities of enhancing crisis management by citizens equipped with sensing technologies. That said, what, in more detail, were the individual motivations in, and the various functions and uses of, producing these radiation participatory monitoring devices and maps?

SAFECAST AND RADIATION WATCH NETWORKS OF PARTICIPATORY RADIATION MONITORING

To understand the individual motivations in, and various functions and uses of, producing these radiation participatory monitoring devices and maps we conducted a questionnaire survey and semi-structured interviews over e-mail with 16 respondents from the two local citizen networks Safecast and Radiation Watch. These networks were highly active in participatory radiation monitoring after the Fukushima Daiichi disaster. We first provide some background on Safecast and on Radiation Watch before presenting our findings.

Safecast is an independent group of approximately 100 enthusiasts with connections to Tokyo and US Hackerspaces, aiming to produce high-precision open data. The participants range from students and professionals from technical fields, to activists, designers, and businessmen. The group began soon after Fukushima to empower people following the problematic communication of the Tokyo Electric Power Company (TEPCO), which ran the

Fukushima Daiichi nuclear power plant, and the Japanese government in the weeks following the disaster. Safecast remains in operation at the time of writing (early 2013), monitoring various areas around Japan, and has extended voluntary monitoring to other countries, for example, measurements were conducted around the San Onofre nuclear plant in California early in February 2012 after a reported minor radiation leak.

The measurement equipment for radiation monitoring ranges from handheld devices to static sensors. Once data are gathered, they are published on the safecast.org website. A staggering 2,523,635 hits were made following the inception of Safecast on March 25, 2011, to February 21, 2012. Safecast uses professional-grade measuring devices and, at the time the survey was conducted, had 85 mobile and handheld devices and 50 static sensors. The project was initially funded by the Kickstarter fundraising website but also acquired funding from other sources, including private investors and universities, including Keio University and its Scanning the Earth Project.[19] We accessed and interviewed Safecast members by way of its mailing list over the course of the fall (August–October) of 2011.

The second group of participants interviewed was recruited from the group Radiation Watch,[20] with approximately 1,200 members. Radiation Watch also aims to empower citizens through implementing radiation measuring as shared technology and *everyday practice.* This enables people to produce data themselves to make radiation monitoring highly accessible collectively. The group, which comprises volunteering engineers, designers and scientists, designed a DIY iPhone accessory to achieve precision similar to standard Geiger counters. The device later evolved into an actual product that no longer needs any DIY skills and can be plugged into Apple devices to measure radiation. Radiation Watch operates a Facebook group for communicating and organizing group members.

To better understand the motivations behind these participants measuring radiation data, how these local practices started, and how the data shaped the opinions of respondents during the measurement process, our questions were organized into five key themes suggested by the situation. We now discuss these themes and our findings.

THEME 1: MOTIVATION TO MEASURE

We first asked about the participants' motivations for undertaking radiation measurements. A key finding was the difference between the aims of Safecast and Radiation Watch members. Safecast members aim to publicize the data to the *general public*, while Radiation Watch members aim to empower *individual citizens* with their own devices to measure radiation. 'Safecasters', as we nominate them, post-Fukushima started measuring radiation as early as March 12, 2011—one day after the disaster struck. Their motivation was to 'make radiation data available to the public', as an alternative to 'government and [other] internet [sources]' (participant 1). In contrast, most Radiation Watch members started relatively late after the disaster, primarily during the second half of 2011. This was partly caused by a gradual development of the design of the DIY iPhone accessory, later offered to the general public. A Radiation Watch member explained: 'I only started measuring when I got a tip on a cheap Geiger counter from a friend of mine during the fall' (participant 7). The 'cheap Geiger counter' refers to the 'Pocket Geiger'; one of the first devices that attracted the attention of a large number of citizens, not only geeks related to Tokyo Hackerspace or Safecast. The easy access technology design of measuring tools and services for sharing data, such as making them work with devices that users were accustomed to, such as iPhones or iPods, facilitated this growth of 'citizen science monitoring'.

Aligning to its focus on the individual, the motivations of Radiation Watch members also tended to relate to personal goals; '[I started to be interested in measuring] when I heard my friends, having three kids, were worrying about radiation. . . . [I] wanted to know if my neighbourhood was safe enough or not. And [I also] wanted to have my own resource to make decisions' (participant 1). Another participant mentioned she started measuring radiation during the summer, around the time of reports on radiation in food: 'I am in Nagoya, far away from Fukushima, but I know that food travels, people travel, gardening soil travels, and I thought it was best to start testing things on my own' (participant 2).

In being asked about their original sources of data prior to starting their own measurements, most Radiation Watch participants responded they had initially relied on official data published by the Ministry of Education, Culture, Sports, Science and Technology (MEXT), as cited in various media, but had found it too technical as well as not providing enough information on an adequate number of locations. One participant noted that prior to engaging with personal radiation monitoring, she had begun searching for alternative information sources such as Safecast and Radiation Watch; the availability of the cheap monitoring devices then motivated her to buy one and participate actively. She wanted to contribute to the general community as much as she wanted to know what the situation was in a given neighborhood for her own activities. The second most common and associated reason for starting to measure related to feelings of insecurity about the radiation spreading within a person's immediate environment. In this case, buying the radiation monitoring equipment and measuring became more of a ritual of receiving some comfort, which we interpret as a fetish function of these DIY tools.

THEME 2: CREATING AND SHARING OPEN DATA

The second question related to identifying with monitoring and sharing radiation data. We presented interviewees with five potential answers to rank on a Likert scale of 1 to 5 where 5 represented 'strongly agree' or 'most important' and 1, 'strongly disagree'. Opinions were very diverse on the issues of 'distrust and protest against government' and 'distrust in TEPCO and media coverage', with regard to radiation data, with replies spread robustly across the scale. This variation we attributed to most people not deciding to start measuring radiation because they mistrusted government data, but because they were interested in better granularity and precision of data, mostly concerning their immediate surroundings and places where they spent most of their time—the house and garden, the playground and the park. A high number of positive answers (strongly agree) on another question on the 'need for more precise data about particular exposure in locations' supported this proposition. One interviewee stated: 'Concern of radiation around my house was the biggest reason. Nobody would come to measure to see if there was a hotspot in my garden' (participant 3). Local hotspots became an important topic in the Japanese media soon after the disaster. The hotspots started appearing at various locations where rainwater flowed, often, and surprisingly, far away from the Fukushima Daiichi reactors and close to Tokyo; with some hotspots discovered by citizen sensing activities. Some participants placed a high importance on the concept of 'open data' and wanted to know more about data collection methods. The expressed belief was that these data were important both to citizens as well as 'governments, companies and researchers', because all could team up to 'benefit (from the data) and tackle the problem in a more cohesive and creative manner' (participant 4).

The impression, therefore, that these responses gave is that measuring radiation was not a type of protest or distrust of government and official authorities carrying out measurements but an effort stimulated by a need to do something, to participate, to be part of the solution, to be useful to the community, to have a feeling of empowerment through participation at a time when personal empowerment was seriously compromised by the scope and gravity of the nuclear disaster, and to contribute to more accurate readings at the personal local level, which was not adequately supplied by official data. These reasons also inform our notion of the fetish function of Geiger counters, as people are aware that they can do little about the effects of radiation but they still like to measure it in the perception they can do something, for example, to avoid hot spots. Overall, the majority of respondents leaned toward becoming more cautious and feeling less comfortable in being in areas of possibly increased radiation that they had not measured. Conversely, they were more comfortable about walking and inhabiting places where they had measured radiation levels to inform their actions. Measuring, gathering and aggregating data and interpreting their meaning thus made people more aware of the effects of radiation. A typical reaction was that 'measuring allows oneself to see what is happening and takes away the mystery and with that most of the fears' (participant 1). Accordingly, this created the impression of empowerment linked to personal safety and comfort for most participants.

THEME 3: EVALUATING THE DATA AND EXPERIENCE WITH MEASURING

The most important benefit that participants mentioned was that measuring and knowing radiation levels in their immediate surroundings created a feeling of control of the situation, a form of empowerment over concerns, such as consuming locally produced food. Many participants also mentioned that knowing more about radiation comforted them, as one respondent stated:

> Even though I am not an "expert", I can say my area is safe enough based on the knowledge I collected and numbers that my DIY machines reported. Additionally, I gained ability to judge if (officially) provided information is correct or not. (participant 1)

THEME 4: ACCURACY OF CITIZEN SCIENCE DATA

The issue of data validity gathered by citizens was strongly recognized by the respondents. One respondent described the attitude of the official scientists and media toward citizen scientists as dismissive, and he felt discouraged because his data was labeled as amateurish and thus questionable (participant 4). But another respondent took this as a reason 'to contribute and make monitoring easier to be used by more people. This included measurement protocol standardization and consolidation of [our] database with other efforts' (participant 10). The post-Fukushima 'citizen scientists' were thus quite aware of the issue of data accuracy and its usefulness, and they considered ways to ensure precision, such as improving measuring devices to automatically log additional metadata about the measurement location by which to compare data from other devices. One participant also expressed disappointment that not only Japanese but also foreign media ignored data collected by citizens: 'I have been following the mainstream media—in English—quite intensively (*Japan Times, Asahi, NHK, Mainichi, Kyodo,* etc.) and do not recall ever having seen an article that includes data collected by citizens. I think it is generally ignored' (participant 2).

A noticeable difference, however, was apparent in the responses on aspects of this topic between Safecast and Radiation Watch participants. One Radiation Watch respondent commented:

> They [traditional researchers] usually deny amateurs and dispute our data, but I just want to know if my place is safe or not (I do not need to know if my place is 0.05uSv or 0.06uSv). In my opinion, they struggle to explain their data in plain Japanese. (participant 1)

This statement emphasized the role of language in constructing the meaning of data for accessible everyday consumption, which was ignored by official measurements, and was another key reason for citizen monitoring efforts. This was because the majority of people did not have any technical education in this field. They struggled to understand the data gathered and issued by the official institutions, which was highly technical and not translated into everyday language.

Safecast respondents, whose focus was more technical and professional, were more certain about the validity of their data and had a more positive view of how their data was being evaluated by the science community and media: 'We have had a positive response from researchers and our system is being used by multiple research teams' (participant 1). The difference in perception of data validity related to the focus of both groups. Safecast was more interested in developing and following rigid methodologies for radiation measurements and using higher grade instruments, while Radiation Watch was more focused on crowdsourcing and involving individuals.

THEME 5: FUTURE CITIZEN SCIENCE

This theme focused on the overall experience of monitoring and probed whether this motivated the participants to be involved in any community projects related to citizen science in the future. More than 75% of respondents said they wanted to continue to measure radiation. A typical response was: 'It is something that will likely influence us in the near future, over the next 30 years or so' (participant 5). Another participant mentioned the safety and well-being of his children as the motivating factor to continue: 'Now I want to know if my place is in the same situation as Fukushima, and what I should do first to save my kids [from health risks in the future]' (participant 1). Yet, another respondent pointed out the potential of other nuclear disasters like Fukushima Daiichi in seismically unstable Japan:

> I will continue to monitor my own vicinity. Since Nagoya is located between Monju and Hamaoka, I think it is advisable to become familiar with how to use a monitoring device and to understand the readings . . . just in case we have a similar accident here (participant 2).

The Hamaoka nuclear power plant has two reactors, constructed in 1971 and 1974 respectively. The reactors sit above a major fault line (on this topic, see also Hara in this book) close to the location of the expected epicenter of the next Tokai earthquake. Earthquakes occur regularly in the Tokai region, with an interval of 100–150 years between quakes and the next one is expected to be of magnitude 8 with a 70% chance of the quake recurring in 2012 or thereabouts (*Wikipedia* n.d.). As such, there has been an ongoing effort by the relevant local government to shut down the plant to minimize the potential of another disaster similar to that of Fukushima Daiichi.

In their involvement in future radiation endeavors, Radiation Watch participants again put personal safety and self-empowerment as their main interest, while Safecast members emphasized developing a large scale independent and alternative measurement network—as one respondent put it: 'The goal is to expand to cover Japan entirely and worldwide and repeat measurements to catch radiation trends' (participant 1).

Another question probed what interviewees saw as the most viable citizen science projects for the future based on their experience with radiation monitoring. All respondents had a positive opinion about the potential of data crowdsourcing and support platforms to share data. One respondent stated: 'I believe that crowdsourcing data collection is the way to go. It is more and more important to collect meaningful data' (participant 6). Another respondent mentioned the importance of collective efforts to achieve bigger goals as the best way for the future: 'This is how the future will be—many small efforts can easily become a big one with the help of technology and open thinking' (participant 1). Another respondent expressed a more radical approach that questioned the public credibility of government and corporate radiation monitoring efforts: 'I think it [civic radiation monitoring] will become more widespread, especially as citizens become more frustrated with the obfuscation of the facts by the government and corporations [like TEPCO]' (participant 2).

The differences expressed between Safecast and Radiation Watch participants on these themes then led us to explore these differences in more depth.

COMPARING SAFECAST AND RADIATION WATCH

While Radiation Watch participants were mainly concerned with individual self-empowerment and their immediate and personal situations, Safecast members developed a more activist and collectivist approach to deal with the broader impact of the Fukushima Daiichi disaster. Instead of focusing on individual areas, Safecast focused on mapping all areas affected. This made Safecast members more motivated to work with multiple stakeholders, including professional ones, while Radiation Watch focused on the data generated by a large number of citizens mainly interested in their own spaces. Arguably, Radiation Watch's more individualized attitude to radiation monitoring and crisis reflects the thrust of 'reflexive modernisation' (Beck 1994), with all its ambiguities of individualization, universality of risk, and homogenization based on safety.

By way of contrast, the participatory monitoring efforts of Safecast placed higher emphasis on collective hacking of hardware and other DIY activities, and attempted to create a 'global' consensus and support for independent data measurements through the Hackerspace network. Safecast thus aimed to empower not only individuals but also communities in a more radical and 'agile' way (Ito 2011). More broadly, Safecast represented an effective 'international' response to a crisis in terms of accessing the needed resources (money and tools) internationally for civic monitoring of radiation. As such, the Safecast movement can be seen to embody the notion of 'cosmopolitical citizenship' as civic action based on designing and building new tools to support various functions rather than the more 'anxious' individual 'reflexive' approach to radiation monitoring.

The participatory monitoring efforts of Geiger counters and similar low-tech solutions in both cases, however, demonstrated the difficulty of getting accurate radiation data and deciding on appropriate courses of action to cope with the Fukushima post–nuclear disaster situation. In this context, participatory monitoring was not only about the crowdsourcing of data and dispersing individual and collective anxieties, hopes and fears, but also about creating a more critical attitude and realistic expectations toward technologies in terms of understanding their limits. This understanding of the DIY technologies enabled citizens to become more agile in their response and expectations (although this was more typical of Safecast users).

Another function of the DIY tools (as more apparent in the case of Radiation Watch) was to empower citizen users at a more private level, where tools also appeared more a type of fetish, as therapeutic devices for a post-apocalyptic ritual of catharsis and healing with elements of personal protest and reflection (Kera 2012). In this sense, the DIY monitoring of radiation tools could be seen to represent modern-day fetishes and power objects with the ability to connect anxiety and hope, provide symbolic and real power over chaotic circumstances, and link scientific data with primal human emotions to build local capacity to better deal or cope with very trying post-apocalyptic circumstances.

CONCLUSION

Participatory monitoring of radiation in Japan after the Fukushima Daiichi nuclear disaster poses a unique case of local and citizen social and technical convergences and synergies, which suggests a resilient model for emergent post-apocalyptic citizenship. This finding is particularly revealed in investigating the synergy between the loose and global network of geeks from Hackerspaces with post–nuclear disaster citizens (or victims) in Tokyo, which led to the creation of 'humanitarian hardware' tools and enabled participatory monitoring of radiation. Tied to this were emergent online platforms for crowdsourcing and crowdfunding, which provided necessary resources for participatory monitoring. This complex, sociotechnical infrastructure of open source hardware tools, online services, cooperation around, and workshops in, the Tokyo Hackerspace evolved very quickly post-disaster.

This dynamic infrastructure, which no official government nor nongovernment actors could have anticipated, enabled citizen volunteers to begin robust grassroots radiation crowdsourcing efforts. These political and decision-making efforts at the grassroots level were always closely linked to processes of designing, building, and testing, prototypes, especially DIY Geiger counters, and special lamps. Participatory monitoring of radiation thus amounts to a form of political and collective action around prototypes, which defines its politics as a form of sociotechnical design experiment. The various functions of the designed tools, from the more pragmatic (to identify radiation hotspots) to the more symbolic (to gain symbolic power in a situation of uncertainty) evolved through these experiments. Rather than defining future collectives in terms of risk, discipline, normalization, or biopolitics, these functions of civic radiation monitoring enabled a 'cosmopolitical' citizenship to emerge in the form of pragmatic and plural collectives around DIY tools with various, even conflicting goals and aspirations, particularly in domains of the private and public. As such, citizen participatory radiation monitoring offered a more resilient and agile 'model' for dealing with the complexity and uncertainty of radiation pollution in Japan post-Fukushima. More broadly, the emergence and significance of DIY tools and open source hardware in enabling a new type of citizen empowerment—cosmopolitical citizenship—as vividly displayed in the case of the Fukushima disaster, poses constructive participatory governance lessons for more effective disaster management in the future.

NOTES

1. See also Campbell et al. (2006), Goldman et al. (2009) and Paulos and Jenkins (2005).
2. Ushahidi is an open source tool to crowdsource information using multiple channels, including SMS, email, Twitter and the web, accessed July 27, 2012. http://www.ushahidi.com/.
3. Crisis Mappers: The Humanitarian Technology Network, accessed July 27, 2012. http://crisismappers.net/.
4. Safecast, accessed July 27, 2012. http://blog.safecast.org.
5. Radiation Watch, accessed July 27, 2012. http://www.radiation-watch.org/.
6. An important role in the convergence of research and activism was played by global networks of individuals involved in Hackerspaces (Kera 2012), as a network of coworking spaces in almost 500 cities worldwide inspiring various types of grassroots 'hacktivist' projects connecting citizen science with some form of monitoring and crowdsourcing of data or hacking tools.
7. Google Maps, accessed July 27, 2012. https://maps.google.com/.
8. Google Earth, accessed July 27, 2012. http://www.google.com/earth/.
9. IBM Many Eyes, accessed July 27, 2012. http://www-958.ibm.com.
10. "Open Source Hardware Definition (OSHW) 1.0," Open Source Hardware, accessed July 27, 2012. http://freedomdefined.org/OSHW.
11. Arduino is a popular open source single-board microcontroller used for the DIY Geiger counters: Arduino, accessed July 27, 2012. http://www.arduino.cc/.
12. Pachube is an online database for sensor data and platform for building applications using sensor data on the environment. Pachube was relaunched under the new name Cosm in May 2012: https://cosm.com/.

13. *Make* magazine, accessed July 27, 2012. http://makezine.com/ and Maker Faire, accessed July 27, 2012. http://makerfaire.com/.
14. Kickstarter, accessed July 27, 2012. http://www.kickstarter.com/.
15. Kickstarter, "RDTN.org: Radiation Detection Hardware Network in Japan," accessed July 27, 2012. http://www.kickstarter.com/projects/1038658656/rdtnorg-radiation-detection-hardware-network-in-ja.
16. Safecast is a global sensor network for collecting and sharing radiation measurements to empower people with data about their environments. Safecast, accessed July 27, 2012. http://blog.safecast.org/.
17. Kickstarter, accessed July 27, 2012. http://www.kickstarter.com/.
18. Japan Geigermap, accessed July 27. 2012. http://japan.failedrobot.com/.
19. Scanning the Earth Project, accessed July 27, 2012. http://scanningtheearth.org.
20. Radiation Watch, accessed July 27, 2012, http://www.radiation-watch.org/.

REFERENCES

Akiba. "Kimono Lantern and Humanitarian Open Source Hardware." *FreakLabs* (blog), 2011, accessed June 10, 2012. http://freaklabs.org/index.php/Blog/Misc/Kimono-Lantern-and-Humanitarian-Open-Source-Hardware.html.

Bäckstrand, K. "Civic Science for Sustainability: Reframing the Role of Experts, Policy-Makers and Citizens in Environmental Governance." *Global Environmental Politics* 3, no. 4 (2003): 24–41.

Beck, U. "The Reinvention of Politics: Towards a Theory of Reflexive Modernisation." In *Reflexive Modernisation,* edited by A. Giddens, S. Lash, and U. Beck. 1–55. Cambridge: Polity Press, 1994.

Campbell, A., S. Eisenman, N. Lane, E. Miluzzo, and R Peterson. "People-centric Urban Sensing." *Proceedings of the 2nd Annual International Workshop on Wireless Internet WICON 06* 12, no. 4 (2006): Article 18.

Gershenfeld, N., R. Krikorian, and D. Cohen. "The Internet of Things." *Scientific American* 291 (2004): 76–81.

Giddens, A. *Modernity and Self-Identity*. Cambridge: Polity, 1991.

Goldman, J., K. Shilton, J. Burke, D. Estrin, M. Hansen, N. Ramanathan, S. Reddy, V. Samanta, M. Srivastava, and R. West "Participatory Sensing: A Citizen-Powered Approach to Illuminating the Patterns that Shape our World." White Paper, 2009, accessed August 29, 2012. http://wilsoncenter.org/sites/default/files/participatory_sensing.pdf.

Goodman, E. "Three Environmental Discourses in Human-Computer Interaction." *Proceedings of the 27th International Conference on Human factors in Computing Systems—CHI EA '09*. New York: ACM Press, 2009.

Greenfield, A. *Everyware: The Dawning Age of Ubiquitous Computing*. Berkley, CA: New Riders Publishing, 2006.

Hedgecock, R. "Mobile Air Quality Monitoring and Web-Based Visualization." MA thesis, Venderbilt University, 2009.

Hooker, B., W. Gaver, and A. Steed. "The Pollution e-Sign." *Design* (2007): 1–4, accessed June 10, 2012. http://itp.nyu.edu/sustainability/interaction/uploads/pollution_e-sign.pdf.

Ito, J. *Innovation in Open Networks* (video). Fora.tv, November 10, 2011, accessed June 10, 2012. http://fora.tv/2011/11/10/Joi_Ito_Innovation_in_Open_Networks.

Japan Today. "Radiation Hotspot in Chiba Linked to Fukushima." *Japan Today,* November 29, 2011, accessed February 19, 2012. http://www.japantoday.com/category/national/view/radiation-hotspot-in-chiba-linked-to-fukushima.

Jasanoff, S. "Technologies of Humility: Citizen Participation in Governing Science." *Minerva,* 41(3), 2003: 223–244.
Kera, D. "Hackerspaces and DIYbio in Asia: Connecting Science and Community with Open Data, Kits and Protocols." *Journal of Peer Production* no. 2 (2012), accessed August 29, 2012. http://peerproduction.net/issues/issue-2/peer-reviewed-papers/.
Latour, B. *We Have Never Been Modern.* Cambridge, MA: Harvard University Press, 1993.
Latour, B. *Politics of Nature: How to Bring the Sciences into Democracy.* Cambridge, MA: Harvard University Press, 2004.
Law, J., and J. Hassard. *Actor Network Theory and After.* Oxford, UK: Wiley-Blackwell, 1999.
Paulos, E., R. Honicky, and B. Hooker. "Citizen Science: Enabling Participatory Sensing." In *Handbook of Research on Urban Informatics: The Practice and Promise of the Real-Time City,* edited by M. Foth, 414–434. Hershey, PA: Information Science Reference, 2009.
Paulos, E., and T. Jenkins. "Urban probes: Encountering our Emerging Urban Atmospheres." *Methods* 7 (2005): 341–350.
Peterova, R. "Urban Interfaces and Extensions: Sensors, Chips, and Ad Hoc Networks as Tools for Urban Culture." MA thesis, Charles University, 2011.
Reuseum. "Update From Akiba@TokyoHackerSpace." 2011, accessed June 10, 2012. http://www.reuseum.com/2011/04/update-from-akiba-tokyo-hackerspace/.
Saavedra, J. "Modular // Neuroid: Wearable, Reconfigurable Personal Pollution Monitor," 2011, accessed June 10, 2012. http://thesis.jmsaavedra.com/papers/.
Weiser, M. "Some Computer Science Issues in Ubiquitous Computing." *Communications of the ACM* 36, no. 7 (1993): 75–84.
Wikipedia, s.v., "Tōkai Earthquakes," accessed July 27, 2012. http://en.wikipedia.org/wiki/T%C5%8Dkai_earthquakes.

7 Envirotechnical Disaster at Fukushima

Nature, Technology and Politics

Sara B. Pritchard

It was a triple disaster. A shallow, magnitude 9.0 earthquake rocked Japan for an astounding six minutes (World Health Organization 2012: 2).[1] The destructive trembler, the largest ever known to have hit the island nation (US Geologic Survey 2011), triggered a huge tsunami that eventually rose to the height of a four-story building (Biello 2011, Perrow 2011, Tabuchi 2011). The foreboding wall of water rushed toward northeastern Japan, barreling down at the speed of a jumbo jet (CTV.ca News Staff 2011), slamming ships into bridges and washing away coastal villages and highways in mere seconds. Then, three of the six nuclear reactors at the Fukushima Daiichi nuclear power station owned by Tokyo Electric Power Company (TEPCO) experienced full meltdowns (CNN Wire Staff 2011). A single word—Fukushima—now stands for the multifaceted complexity of the events which took place on March 11, 2011, and all that has transpired in the months ever since.[2] Moreover, the temporal implications of the triple disaster gained new meaning when Japan's atomic commission announced in October 2011 that it may take more than three decades to clean up Fukushima (McCurry 2011).

Even before Fukushima, the early 21st century had already offered scholars in fields like science science, technology and society studies (STS) and environmental history significant research and teaching moments.[3] Now Japan's triple disaster provides yet another occasion for us to engage with pressing questions—questions about the construction, maintenance, and future mix of energy regimes, both politically and technologically, in the modern world (Hecht 2009, Mitchell 2009, 2011); by association, the development, expansion, and implications of the atomic age for both humans and nonhumans (Bocking 1995, Hamblin 2008); and the relationship between nature (including human bodies) and technological systems (Hughes 2004, Parr 2010, Pritchard 2011).[4] Indeed, all of these issues are fundamental to ongoing debates about how to create a more sustainable world.

Several leading scholars, including sociologist Charles Perrow and historian Thomas Parke Hughes, have studied the design and operation of large-scale, modern technological systems like those at Fukushima Daiichi.[5] Building on their insights, in this chapter I examine the Fukushima Daiichi accident as an *envirotechnical* disaster, a result of the convergence of natural and sociotechnical processes. I argue that the concept of envirotechnical systems is a valuable way to understand what happened that also goes beyond what Perrow and Hughes offer through their concepts of normal accidents and technological systems, respectively. In the final section, I use the notion of envirotechnical regimes to stress the strategic configuration of Fukushima Daiichi's envirotechnical system, highlighting the ways in which political and economic power shaped the making of the facility, both during normal operations and throughout the events that began to unfold on March 11, 2011.

My wider aim in this chapter is to consider how both actors and scholars think about the relationship between nature and technology, in part because such categories, categorizations, and relationships matter in the so-called real world. For instance, assumptions about the relationship between environmental and technical systems shape the design and operation of 'technological' systems like nuclear reactors. Alternative understandings like envirotechnical analysis (Pritchard 2011: 11–24), might therefore better inform and thus ultimately enrich contemporary discussions of energy technologies and the development of energy policy, not only in Japan but globally.

NORMAL ACCIDENTS, TECHNOLOGICAL SYSTEMS, AND ENVIROTECHNICAL ANALYSIS

In 1984, Charles Perrow published *Normal Accidents*, his now-classic study of the partial meltdown at Three Mile Island five years earlier, arguing that complex, tightly coupled systems invariably lead to accidents (see Perrow 2007). In his analysis, Perrow highlighted the unpredictable dynamics of sociotechnical systems (Bijker et al. 1987, Hughes 1993: 6) given their size, complexity, and inextricability. Journalists from the *New York Times* expressed a version of Perrow's idea two days into the Fukushima crisis when they described 'a cascade of accumulating problems' (Sanger and Wald 2011).

Perrow asserted, however, that it is misleading, if not hazardous (Marx 2010), to use the common term 'accidents' to describe situations like that of Fukushima Daiichi because it minimizes the inherent risks of modern technological systems. Such language implies that accidents are caused by technical glitches or human error, when instead they should be understood as intrinsic to those systems. In short, accidents are normal and systemic, not extraordinary and inadvertent. Overall, Perrow's work challenges reductive thinking and has enabled scholars, not to mention technologists and policy makers, to perceive systemic vulnerabilities (e.g. Bijker 2009) within complex technologies like nuclear reactors.

Nonetheless, in the aftermath of disasters like Chernobyl, Katrina, and now Fukushima, government regulators and industry officials often focus on trying to 'fix' the technology in question and reduce the likelihood of future human error. Such a techno-fix mentality conveniently focuses on supposedly mere technical problems (e.g. Rosner 2004), reifies the perceived division between the social and the technological, and ignores the fact that how problems get framed shapes what solutions are possible and thus actionable. Consequently, people involved in such debates are less likely to ask deeper, far more difficult questions about those technologies—questions such as: Whose goals do these technologies serve? What political and economic interests shape the design and use of technological systems? And what assumptions about the natural world and human-natural relations are embedded in and shaped by these technologies?

Also in the mid-1980s, Thomas Parke Hughes theorized the concept of technological systems within the context of the history of technology. In his book *Networks of Power* (1993), Hughes traced the development of electricity networks in the United States and Western Europe. By focusing on systems, Hughes showed how technological artifacts rarely exist in isolation. Moreover, system builders create not just technical infrastructure like power lines but also the capital, market demand, political support, and values that enable and perpetuate that system. Thus, Hughes reminds us—scholars, technical experts and citizens alike—to take heed of the whole. Back-up generators, for instance, may seem like an old, even mundane technology (Edgerton 2007), especially in the supposed advanced West; but as the crisis at Fukushima Daiichi made abundantly clear, they are essential to the safe operation and shutdown of nuclear power stations.

Hughes's understanding of technological systems was predicated, in part, upon an explicit conceptualization of the relationship between the environment and technology. As Hughes declared, likely echoing the mindset of the systems theorists he studied (Hughes and Hughes 2000):

> Those parts of the world that are not subject to a system's control, but that influence the system, are called the environment. A sector of the environment can be incorporated into a system by bringing it under system control. An open system is one that is subject to influences from the environment; a closed system is its own sweet beast, and the final state can be predicted from the initial condition and the internal dynamic. (Hughes 1993: 6)

By conceptualizing technological systems in this way, Hughes suggested that technology largely shapes the environment; only in 'open' systems does the environment shape technology.[6]

Contrast, however, Hughes's framework with Perrow's less-known concept of 'eco-system' (not to be confused with 'ecosystem') and specifically of an 'eco-system accident'. As Perrow put it, an eco-system accident is the result of 'an interaction of systems that were thought to be independent but are not because of the larger ecology'. More precisely, such accidents 'illustrate the tight coupling between human-made systems and natural systems. There are few or no deliberate buffers inserted between the two systems because the designers never expected them to be connected' (Perrow 1999: 14, 296). According to Perrow, eco-system accidents result precisely from the inability to realize the high-modernist ideal of neatly cleaved technological and environmental systems. Given what has already taken place in Japan, not to mention what will likely unfold in the years ahead, Perrow's eco-system probably resonates with us more than Hughes's technological system. Indeed, in light of 3/11, Perrow seems painfully prophetic.[7]

At the same time, scholarship at the intersection of the history of technology and environmental history both extends and refines the contributions that Perrow and Hughes offer us. Over the past two decades, those working at the nexus of these fields have developed historical and theoretical insights, which have been synthesized in several essays (Gorman and Mendelsohn 2010, Pritchard 2011: 11–24, Reuss and Cutcliffe 2010: 1–6, 291–301, Stine and Tarr 1998). Some scholars have focused on historical actors' ideas about nature, technology, and their relationship, sometimes tracing how these cultural attitudes then shaped interactions with the material world. Particularly crucial is the strategic definition of these terms. Other scholars have explored how people have used the environment, particularly in managing nature as essentially technology to do things—from developing chicken breeds to maximize agro-industrial production (Boyd 2001), to breeding lab-ready hemophiliac dogs (Pemberton 2004). Edmund Russell (2004) calls them biotechnologies. Such examples suggest how humans have interacted with and ultimately transformed nonhuman nature, thus further challenging the notion of a supposedly pristine nature without human influence (Cronon 1995). Still others have argued that environmental factors and natural processes shape (but do not determine) technological change, an assertion that challenges recent accounts of technical development within the history of technology that highlight the social, political, and cultural shaping of technology (Bijker et al. 1987). Envirotech scholarship, which has been influenced by STS work on hybridity (Haraway 1991: 149–181, 2003, 2008, Latour 1993), actor networks (Callon 1987, Latour 2005, Law 1987), and coproduction (Jasanoff 2004, Kline 2000), thus seeks to explore dynamic relationships between nature and technology—physically and culturally, historically and analytically.

Integrating the contributions of this literature with the insights of Perrow and Hughes better explains what took place at Fukushima and why, in at least three ways. First, envirotech scholars push us to see the environment as *always* being part of technological systems, not just, as Hughes asserted, of open systems. The concept of *envirotechnical systems* (Pritchard 2007, 2011) encapsulates and foregrounds this dynamic imbrication of natural and technological systems.[8] As the term itself suggests, we might think of these systems as mutually articulating. Yet, as Fukushima demonstrated, their choreography is not necessarily synchronized or even synchroniz*able* (Edwards 2011).[9] The concept of *envirotechnical* system captures this entanglement and dialectical shaping.

Nevertheless, these entanglements of environmental and technological systems do not emerge out of thin air. Rather, they arise from particular historical, cultural, and political contexts. Thus, a related concept, *envirotechnical regime* (Pritchard 2007, 2011), stresses the historical and political production of envirotechnical systems like those found at the Fukushima Daiichi nuclear power plant. This concept emphasizes the specific, often strategic articulation of natural and technological systems to serve particular ends, although, as the disaster showed, these configurations do not always develop exactly as the people and institutions promoting them intended. Finally, envirotech perspectives also call attention to how different groups thought about the definition, relationship, and dynamics between natural and technical systems, often quite strategically. In the wake of Fukushima, for example, industry and government officials have conveniently pointed to the earthquake and tsunami to absolve them of responsibility. Together, these insights complicate and enrich Hughes's understanding of technological systems, in the process lending more credence to Perrow's concepts of normal and eco-system accidents.

INTERPRETING 3/11

On one level, Fukushima Daiichi's reactors were envirotechnical by design. Radioactive elements fueled nuclear chain reactions that eventually generated electricity. Reactors also incorporated, to borrow Hughes's phrase, water to regulate cooling processes in their cores, as well as the storage ponds housing spent fuel rods (Sanger and Wald 2011). From the outset, these entities were both natural and technological in that they were mobilized for their valuable properties and yet also managed for specific purposes—from producing energy to diffusing heat (see also Falk in this book). In addition, almost two decades ago, Arthur McEvoy (1995) encouraged environmental historians to consider the bodies of workers—in this case, largely short-term laborers subject to dubious labor, health, and safety standards (Hecht 2012)—as part of the 'nature' of industry. While some environments were integral to the functioning and operation of the reactors, others were kept out. At least this was the goal. Engineers designed the facility to withstand a tsunami 10.5 feet high. Because the complex was located on a cliff 13 feet above the Pacific Ocean, they believed that Fukushima Daiichi would be safe. But on 3/11, the tsunami was almost twice the height of the facility and cliff combined (Perrow 2011). Furthermore, the massive earthquake caused land subsidence from one to three feet in northeastern Japan, making the region even more vulnerable to the ensuing tsunami (Geospatial Information Authority of Japan 2011).

During the actual crisis, government officials, plant managers, and workers both tightened and transformed the imbrication of the natural and the technological at the Fukushima Daiichi the power plant. We can see these processes at work by examining, in turn, water, air, and the bodies of workers at the complex; although it is worth noting that these terms do not adequately capture their hybrid forms as both natural and technological entities at Fukushima Daiichi. Water was at the center of emergency measures during the crisis's initial phase. As the first 48 hours of the disaster unfolded, our television and computer screens were filled with images of standard firefighting equipment—trucks, hoses, and diffuse spray—all of which was dwarfed by the facility. The fear was that failing to restore normal cooling processes or to establish emergency measures to replace them—and quickly—might lead to a catastrophe. As reporters from the *New York Times* explained in rather dry language on March 13: 'A partial meltdown can occur when radioactive fuel rods, which normally are covered in water, remain partially uncovered for too long. The more the fuel is exposed, the closer the reactor comes to a full meltdown' (Tabuchi and Wald March 13, 2011).

Indeed, water levels inside the reactor cores had already begun to fall. Estimates varied during the first few days, in part because key gauges were not working. However, government and industry specialists guessed that the top four to nine feet had been exposed to air, giving rise to the risk of a partial and possibly full meltdown of the cores. To compound matters, the cooling ponds for spent fuel rods, recently discharged from the reactor cores and thus highly radioactive, also incited concern. Two days into the disaster, experts had already expressed fears that some of these rods had become exposed to air and begun emitting gamma radiation, the most lethal form of radiation to people (Parr 2010: 66–67, Sanger and Wald 2011, Schmid 2011). Indeed, when TEPCO workers were finally able to fix a gauge in reactor number 1 in May 2011, it showed that the water level in the reactor was lower than expected, despite the massive infusion of seawater. In fact, 'one of the most startling findings announced Thursday was that water levels in the reactor vessel, which houses the fuel rods, appeared to be about three feet below where the bottom of the fuel rods would normally stand' (Tabuchi and Wald May 12, 2011). The material properties and qualities of water from the surrounding natural environment, initially external to the facility, were thus vital to the safe operation of the complex, ostensibly high-tech system. The emerging crisis highlighted, indeed magnified, this dependency.

Part of the problem facing those trying to control the situation at Fukushima Daiichi is that nuclear reactors are never—and can never be—completely turned off. A chain reaction may be stopped and the reactor is, at least in theory, safely shut down. Plant managers and technicians did have time to perform these protocols before the reactor cores started melting. This was a major difference from Chernobyl, which also lacked a hard containment shell. However, residual heat in the reactors remains for two reasons. The reactors had been operating at high temperatures (550 degrees Fahrenheit) that resist dissipating very quickly. More significantly, even once the facility has technically been turned off, the fuel still produces heat due to continuing radioactivity—the release of subatomic particles and of gamma rays. It may be only 6% of the heat produced during normal operations but that is still plenty. Pumps must therefore keep water circulating through the reactor core and spent fuel storage ponds, and, crucially, the temperature of that water must be closely regulated by pulling warmed water to a heat exchanger and bringing in cool water to draw off that heat. Otherwise, the cooling fluid will evaporate—and quickly. Recalling Perrow, a cascade effect threatens to make a bad situation even worse: as radioactive decay continues, more heat is produced, which boils off more water, causing water levels to drop further, exposing more fuel to steam and air, which results in greater fuel damage, raising temperatures even higher, which causes still more water to evaporate, and so on. This downward spiral only increases the possibility of a meltdown. Without electrical power at Fukushima Daiichi, customary cooling processes were inoperable and thus ineffective, thereby precipitating precisely the kind of scenario described above (Sanger and Wald 2011).

However, although the plant's cooling system and back-up support had failed, operators perceived the nearby Pacific Ocean and the air surrounding Fukushima Daiichi as part of the emergency control system. Or rather, they would *become* part of that system, indeed vital to it. Consequently, technicians developed a makeshift practice of flooding the reactors' containment vessels with seawater and letting the fuel cool by boiling off that water. However, as that water boiled, pressure in the vessel increased and actually became too high to inject more water. As one American official explained, forcing water into the vessel was like 'trying to pour water into an inflated balloon'. They therefore had 'to vent the vessel to the atmosphere, and feed in more water, a procedure known as "feed and bleed" ' (Sanger and Wald 2011). The key was to ensure that plant operators kept an adequate supply of water flowing into the containment vessels to make up for the 'lost' water as it heated, turned into steam, and was vented. Instrument problems compounded an already difficult situation because workers were not sure how much water remained in the reactors and therefore how much water needed to be injected. In effect, they were 'flooding blind' (again, also see Falk in this book). Moreover, the emergency solution was not even a short-term fix. Nuclear engineers estimated that the process of injecting water could entail several thousands of gallons per day, for 'potentially as long as a year' (Fountain 2011, Sanger and Wald 2011, Tabuchi and Wald March 12, 2011, Tabuchi and Wald March 13, 2011). After all, given the radioactive elements' continuing decay (not to mention their long half-lives), the process of injecting water, followed by its warming, evaporation, and sanctioned release, would necessarily be ongoing. In retrospect, we know that these efforts, heroic but piecemeal, did not prevent the meltdown of the three reactor cores.

These emergency practices created new relationships between ecological and technological entities within and outside Fukushima Daiichi, demonstrating that the boundaries of these systems were fluid, dynamic, and negotiable. Some of the ensuing relationships were inadvertent because no one had planned the earthquake, tsunami, or reactor problems at Fukushima Daiichi (let alone all three) in the first place. Nonetheless, the tsunami had blurred the borders between the hydraulic and atomic. However, other relationships were intended to solve or at least diminish the severity of the disaster precisely because plant operators and nuclear regulators perceived environmental and technological boundaries as negotiable and sought to make them even more permeable as part of their crisis management efforts. The 'feed and bleed' procedure is a prime example of the ways in which borders broke down because some groups perceived them as porous and acted accordingly. Yet, in the process, desperate officials and workers ended up introducing new 'wrinkles' in the 'cascade of accumulating problems' at Fukushima Daiichi.

The influx of seawater from the Pacific, later laced with boric acid to prevent the fuel from reaching criticality, attempted to contain a situation that seemed to be spiraling out of control (Behr 2011). That fluid may have averted even more disaster, but this solution ended up creating a new problem: the liquid, once savior, had now itself become dangerous, a risk object (Hilgartner 1992). Not all of the doctored seawater injected into the reactors boiled off (and thankfully so). Consequently, what remained became contaminated with radiation as well. By June 2011, more than 100,000 metric tons of 'water'—in reality, a salty noxious blend of fresh water, seawater, and radioactive materials that embodied the new envirotechnical system at Fukushima Daiichi—had collected in the bowels of the stricken reactors. One journalist (Biello 2011) called the brew 'a radioactive *onsen* (hot bath)', an erroneous translation of the Japanese word for natural hot springs.[10] Moreover, even in June 2011 (three months after the accident), an additional 500 metric tons of seawater was being poured into the facility every day because leaks prevented normal cooling systems from being restored.

To contend with this literal overflow of contaminated liquid, TEPCO installed devices to filter radioactive residue from the infusion. The utility did so for two reasons: to reduce the volume of contaminated fluid overall and to decrease the amount of radiation so that some of that fluid could be reused on the spent fuel rods (Biello 2011). After all, these rods still needed to be cooled, but without being filtered first, the infusion would leave radiation levels in the reactors too high for workers to continue mitigation efforts. Importantly, this precaution suggests not only that human bodies were becoming envirotechnical objects, but also that those involved in mitigating the disaster at Fukushima perceived them as such, crucial points to which I will return. However, a trial run with the filters was aborted after less than five hours when it captured as much cesium 137 as they expected to be collected in an entire month. In addition, the capacity of tanks TEPCO delivered to store some of the excess fluid proved entirely inadequate, especially as workers continue to spray the reactors daily. Consequently, TEPCO decided to release more than 11,000 metric tons of the toxic brew into the

Pacific Ocean in April 2011. The rationale? Dumping less contaminated water allowed limited storage facilities to be dedicated to more highly contaminated water. Only two months later, additional sanctioned releases were scheduled (Biello 2011, Marran 2011, Monahan 2011). If the tsunami had confounded the boundaries of the sea and reactors, oceanic and the atomic, then cleaning up the accident further blurred those boundaries.

Other boundaries were also transgressed at the nadir of the calamity: just as the water of the Pacific became integral to the attempt to manage the crisis, so too did the atmosphere surrounding the facility. Once doctored seawater was injected and began boiling, plant operators planned releases of steam to reduce pressure in the reactors. The aim was to ease the infusion of more fluid, while simultaneously reducing the likelihood of a catastrophic explosion that might rupture containment shells and release higher doses of radiation from the reactor cores and spent fuel storage ponds into the atmosphere. Radioactive contamination within the reactors meant, however, that this vapor was contaminated as well. Initially, when the fuel was intact, the steam was infused by 'modest' amounts of radioactive materials 'in a non-troublesome form'. However, as conditions within the reactors deteriorated, that steam became 'dirtier' (Sanger and Wald 2011). The planned releases were thus a Faustian bargain (Marran 2011). Without them the situation might well have worsened. Yet, they were apparently not enough. Explosions rocked four of the reactors, indicating that experts did not have as much control over the process of venting excess pressure as they had hoped (Behr 2011, Onishi and Fackler 2011). These episodes undermined confidence in system control, suggesting that large-scale, modern technological systems were fundamentally more vulnerable than many believed or represented (Bijker 2009).

I have focused thus far on the complex dynamics of radioactive elements, water, and air in and beyond the reactors. But we must not forget the human element—from the so-called Fukushima Fifty, a core set of workers who remained at the facility as the crisis deepened in a desperate attempt to regain control over the precipitous cascade of events, to the over 18,000 men who had participated in clean-up efforts by early December 2011 (Hecht 2012). These workers were vital to the system during the crisis and its continuing aftermath. Their labor, their physical bodies, which attempted to assess and repair, feed and bleed in precariously precise proportions, were crucial to getting the new envirotechnical system of Fukushima Daiichi under control, indeed modifying it to reduce the level of risk. However, the extent of their radiation exposure is yet unknown and will probably be debated, if not hotly contested. Given their extended proximity to the power plant, these workers will likely embody, quite literally, new configurations of the natural and technological in the latest chapter of Japan's atomic age.

It is vital to note that TEPCO and nuclear regulatory officials themselves perceived these connections, both as the disaster unfolded and as the herculean clean-up efforts have begun. Four days into the crisis, TEPCO's president asked Japan's Prime Minister Naoto Kan to allow the utility to pull the remaining workers, fearing these men would be exposed to extremely high doses of radiation if they remained at Fukushima Daiichi. Kan refused and installed a trusted aide at the utility's headquarters the next morning, suggesting that the greater good required individual sacrifice (Onishi and Fackler 2011). From the earliest days of the crisis, then, TEPCO constructed workers' bodies as a site where the atomic and biological were coming together dangerously. Conveniently, this allowed the company to claim that it was worried about the health and safety of these 'nuclear heroes', a concern not apparently held for workers during the facility's normal operation and maintenance.

Thousands of clean-up workers are now part of the new system at Fukushima Daiichi. However, these temporary laborers are let go from their jobs once they reach their radiation exposure limit (Hecht 2012), a practice that again suggests how some perceive—and fear—the merging of the natural and the technological in the workers' own bodies. Structuring labor in this way is thus not only a political and economic act, but also an envirotechnical one. Unfortunately, then, these workers have a strong financial incentive to forget the dosimeters that measure their radiation exposure during a given shift to prolong employment, thus increasing their levels of exposure. Their bodies suffer the consequences.

POWER AND POLITICS AT FUKUSHIMA

In this chapter, I not only emphasize the 'nature' of the technological system at Fukushima Daiichi but also argue that we should think of these systems as 'envirotechnical' to represent the ongoing ways that environmental processes both shape and are shaped by technologies. Doing so helps remind us that technologies often obscure our connections to the environment, but 'there are no technologies that remove us from nature' (White 1995: 182).

However, there is a real risk in naturalizing 3/11, paying too much attention to environmental factors and processes involved either in the normal functioning of the reactors or during the crisis. Over a decade ago, Ted Steinberg stressed the 'unnatural' history of so-called natural disasters (2000), hinting at the politics of such categories and categorizations. Unfortunately, Hurricane Katrina and the uneven effects of earthquakes in places like Haiti and Chile have only reconfirmed Steinberg's insights.

Indeed, some Japanese citizens today are wary of analysis that stresses the role of nonhuman nature in technological systems, or anything resembling what Brett Walker calls hybrid causation (2010: 16–20). This was emphasized by Tyson Vaughan, a PhD student in Cornell's Department of Science and Technology Studies, who was recently discussing Walker's concept, as well as envirotechnical analysis, with several people in Japan involved in ongoing debates over mercury poisoning at Minamata Bay during the 1950s. This discussion arose in the context of Vaughan's own research on postdisaster reconstruction efforts in Japan and the United States. His interviewees included the curator of the Minamata Disease Municipal Museum, the curator of Soshisha ('The Supporting Center for Minamata Disease'), a *Nippon Hōsō Kyōkai* (Japan Broadcasting Corporation) journalist, and *kataribe* (storyteller) Miyako Kawamoto, the widow of a leader of one of the main victim advocacy groups.[11] Vaughan conveyed to me that all four interviewees, but particularly Miyako, were distressed by the notion of hybrid causation or seeing technologies as envirotechnical. In their view, focusing, say, on the properties of mercury or how different organisms, including humans, absorbed it at Minamata diverts attention from the corporate and government decisions that contributed to the poisoning of 2,000 victims, with an untold number of others not officially recognized. In short, they fear that multicausal accounts based on complex understandings of historical agency that decenter people as primary causal agents, threaten to diffuse, if not undermine, the responsibility of powerful groups.

It may be tempting for scholars to dismiss their understanding of historical causality, arguing that it is simplistic and fails to reflect scholarly advances. But the past does matter to Japanese citizens today, and it informs their views of nature, technology, politics, and their relationships with one another. For one, other tragedies, including Hiroshima, Nagasaki, and Minamata, loom large in the nation's memory, especially for those active in citizens' movements and the litigation related to these crises. Given such a past, it is understandable why they may be reading Fukushima partly through the lenses of events like Minamata. Indeed, the protracted court battle over the Minamata disaster only concluded on March 22, 2011, 11 days after the situation at Fukushima Daiichi began (Marran 2011). In addition, as Julia Adeney Thomas (2002) has argued, naturalizing state power in 20th-century Japan has historically served to disempower citizens, ultimately crushing social movements and disabling antistate reform. Thus, because naturalization and hybridity have essentially functioned as tools of the state, the response of Vaughan's interviewees to scholarly trends that develop such concepts is even less surprising.[12] After Fukushima, they may be more dubious than ever about framing technological systems as envirotechnical. It seems, at least at first glance, too apolitical: pointing to environmental factors seems to naturalize a crisis while technical explanations carry the false sense of objectivity, both of which evade questions of power and responsibility. Importantly, then, such critiques and the wider political stakes they signal remind scholars of the real world implications of their work.

Nonetheless, I suggest that we can and should read what took place at Fukushima Daiichi as an envirotechnical disaster. Indeed, thinking envirotechnically does not necessarily minimize or obscure the vital political issues these Japanese commentators stress. To the contrary, as Douglas Weiner succinctly put it: 'Every environmental story is a story about power' (Weiner 2005: 409). Indeed, politics are central to the Fukushima of today, in part because they are central to its history. They were inscribed into the facility from the outset, starting with Japan's decision to develop atomic energy (see also Hara, and Juraku, in this book). Facing pressure from the antinuclear movement, industry supporters, not just in Japan, invoked the mantra of safety time and again (Hamblin 2012). Yet, as one former plant operator stated after the crisis began to unfold at Fukushima Daiichi: 'You can take all kinds of possible situations into consideration, but something "beyond imagination" is bound to take place, like the March 11 tsunami' (Yomiuri Shimbun 2011). As the former plant operator suggested, probabilistic thinking, which downplays possibilities like a magnitude 9.0 earthquake or a 14-meter tsunami (let alone both), dominates the nuclear industry, although it is certainly not unique. This tendency is further suggested by the decision of Japan's Nuclear Safety Commission (the equivalent of the US Nuclear Regulatory Commission) not to include any measures regarding tsunamis in its guidelines until 2006, decades after many of the country's reactors were actually built and in operation (Onishi and Fackler 2011, Perrow 2011).

Perrow argued in an essay published shortly after 3/11 that we should instead 'consider a worst-case approach to risk: the "possibilistic" approach', a concept formulated by sociologist Lee Clark (Perrow 2011). There are, of course, powerful vested interests against adopting such an approach. For one, the ways in which notions of the nation and the nuclear have become entangled in the post-1945 world (Hecht 2009), including in Japan, helps to explain past decisions such as the nuclear industry's safety myth and inadequate government regulation of TEPCO.[13] In addition, economics are a powerful form of politics. One TEPCO engineer admitted that he falsified records regarding the containment vessel of reactor number 4 at the Fukushima Daiichi power plant, and design decisions may have prioritized cost and convenience over safety (Perrow 2011).

Politics also shaped how the natural and the technological became entangled at Fukushima Daiichi, whether before 3/11 or afterward. The concept of envirotechnical regime foregrounds the politics of this process, including how specific groups and institutions pushed for linking nature and technology in particular ways, both in situations of normalcy and those of crisis. Let me return to three examples that I have discussed in this chapter.

First, Japan established and expanded a nuclear industry on the edge of the Pacific Ring of Fire, a perimeter known for its major earthquakes and potential for destructive tsunamis. Imperatives of nation-building, industrialization, and modernization drove such decisions in spite of the hazardous environment in which they were being undertaken. In the context of such powerful political and economic motives, supporters of nuclear power in Japan conveniently differentiated natural and technological systems, even as these material linkages were actually being forged and strengthened on the ground. Japan's decision therefore raises vital questions about the politics of siting nuclear power stations and the ways that contexts are analyzed and ignored (see especially Juraku in this book).

Second, the timing of the seawater injections at Fukushima Daiichi has been much debated. TEPCO's administration probably delayed flooding the reactors because doing so 'amounted to sacrificing' them (Behr 2011). According to Akira Omoto, a former TEPCO executive and member of the Japan Atomic Energy Commission, the utility 'hesitated because it tried to protect its assets' (Shirouzu et al. 2011; see also Bradsher et al. 2011, Onishi and Fackler 2011). These seawater injections may indeed illustrate the reblending of environmental and technological systems at Fukushima Daiichi, but the context, timing, and meaning of that process matter as much as the act itself. In other words, it is not just that the oceanic and atomic were again linked through the emergency protocol but that they were connected only when TEPCO, pressured by government leaders, was forced to recognize how disastrous the situation had become.

Finally, the Fukushima Fifty have received a great deal of attention, being heralded in the media for their 'selfless sacrifice'. Yet many more labor below the radar screen—or, more aptly, the dosimeter. However, it is not just a question of the forgotten masses, focusing on the so-termed nuclear heroes at the nadir of the crisis while ignoring the thousands of short-term workers carrying out cleanup as the disaster fades from our memories—at least for those of us located thousands of miles away. Rather, the political and economic context of these laborers, their work, and their bodies is noteworthy. As Gabrielle Hecht (2012) showed, the structure of subcontracting work in the nuclear industry—in Japan but also elsewhere—is such that these workers, often called 'nuclear nomads', are basically left unprotected by environmental health and safety regulations, and unaccounted for in corresponding statistics. Furthermore, most of them are poor day laborers who live in slums surrounding Japanese cities or men from the evacuation zone surrounding Fukushima Daiichi who are desperate for work. This practice conveniently elevates the safety ratings of a given utility like TEPCO or that of the nuclear industry overall while reducing official statistics regarding human health exposure. Thus, it is not just any bodies that are being exposed to increased rates of radiation in Japan, before and after Fukushima; rather, some of the poorest, most economically vulnerable people in Japan are more likely to be affected (and affected more significantly) by the merging of the biological and the atomic in the modern world. Radiation may indeed illustrate the permeability of political borders (Marran 2011), and thereby heighten the regional implications of risk. Yet Fukushima also demonstrates how such crises not only reproduce, but also exacerbate, social inequalities.

As these examples illustrate, the political shaping of Fukushima Daiichi's 'technological' system is unquestionably vital. It demands careful consideration—by scholars and citizens alike. However, politics alone do not explain the triple disaster. Obviously, the earthquake and tsunami have a bearing. As do the properties of radioactive materials, water, air, and human bodies, and the relationships among them both during normal operations and normal accidents. At the same time, as the skepticism of concerned Japanese citizens reminds us, it is indeed hazardous to focus on the 'nature' of the disaster alone, isolating it from the larger system of which it was an integral part. It is precisely the complex, dynamic, porous, and inextricable configuration of the environment, technology, and politics that *together* helps us understand all that the single word 'Fukushima' now signifies (Mitchell 2002, Sutter 2011, Walker 2010). It was due to an earthquake, nuclear reactors, *and* delayed seawater injections; a tsunami, back-up generators located in basements, *and* probabilistic thinking; and continued radioactive decay, spent fuel rods, *and* weak government oversight of industries defined by large-scale, high-modernist technologies. As Michelle Murphy (2006: 180) states in a quite different context, '*And . . . And . . . And . . .*'

CONCLUSION

In the end, Fukushima explodes, so to speak, Hughes's idealized representation of closed and open technological systems. Instead, by thinking in terms of envirotechnical systems, we can avoid the pitfalls of such tidy categories, firm borders, and static notions of both nature and technology that actually fall apart in places like Fukushima Daiichi. Meanwhile, the concept of envirotechnical regimes calls attention to the political processes by which these systems are brought into existence and maintained. This approach therefore encourages us to consider *how, why,* and *in what particular ways* technological and natural systems converge and interact with one another, even when, as Perrow (1999: 296) put it, 'designers never expected them to be connected'. Examples abound, especially in the early 21st century, but Fukushima has particularly—and poignantly—brought these insights home. Indeed, growing concerns about global sustainability, especially around climate change and energy policy, bring new urgency to contemporary debates over the place of nuclear power in the future. What took place at Fukushima Daiichi suggests that there will be no easy answers. However, continued confidence in the ideal of wholly discrete environmental and technological systems is an unfortunate, even dangerous illusion that can only entrench high-modernist hubris and the belief that humanity and technology are separate from the environment.

NOTES

A previous version of this chapter appeared in *Environmental History* 17 (2012): 1–25, published by Oxford University Press on behalf of the American Society for Environmental History and the Forest History Society. I thank the journal and its editor, Nancy Langston, for allowing an updated version to be republished here. I would also like to thank Richard Hindmarsh, Ian Miller, Julia Adeney Thomas, Brett Walker, and two anonymous journal reviewers for their comments on earlier versions of this essay. Finally, I thank Nancy Langston for soliciting the original piece, Tyson Vaughan for his research contributions, and Connie Hsu Swenson for her editorial assistance.

1. For a historical study of earthquakes and their relationship to the nation in Japan, see Clancey (2006).
2. The other shorthand, 3/11, obviously evokes the political valence of 9/11 in the United States while forging a parallel between the two crises and societies. See Anderson (2011).
3. On Hurricane Katrina, see Kelman (2006, 2007), and the *Social Studies of Science* special issue on Hurricane Katrina (Volume 37, February 2007). On hydraulic fracturing, see Tarr (2009). On BP's Deepwater Horizon, see Jones (2010) and Shulman (2010).
4. Scholars of Japan from diverse disciplines (and nationalities) have contributed to the scrutiny of Fukushima since March 11, 2011. Many rich analyses are available through *The Asia-Pacific Journal*'s forum (http://www.japanfocus.org/Japans-3.11-Earthquake-Tsunami-Atomic-Meltdown), the Teach 3.11 website (http://teach311.wordpress.com/), and the '3.11 Virtual Conference: Looking Back to Look Forward' (http://fukushimaforum.wordpress.com/conferences/).
5. I do not take up the question what makes technological systems 'modern' (or not) here. See Latour (1993).
6. One can ask if everything comes under Hughes's definition of environment here, from unstable commodity prices to striking workers, as they too resist system control.
7. For more detailed analyses of Hughes and Perrow, see Pritchard (2012).
8. See also Gardner (2009), Hughes (2004), LeCain (2009), and Reuss and Cutcliffe (2010).
9. For more on these issues, including differences among envirotechnical systems, Perrow's eco-system, and Hughes's later 'eco-technological environment', see Pritchard (2012), especially the notes.
10. Biello's naturalization of the toxic liquid is unfortunate, given that government and corporate interests have worked to focus attention on the earthquake and tsunami. Nonetheless, in view of the longstanding importance of bathing rituals and hot springs in Japanese culture, including Shinto religious traditions, the contaminated fluids within and beyond Fukushima Daiichi are likely freighted with meaning, from offending animistic spirits to possibly being hell on earth. Interestingly, however, the boundaries between baths and natural hot springs are increasingly unclear because these springs are often highly managed landscapes, which would further complicate interpretations of Fukushima Daiichi's fluids as a toxic *onsen*. I thank John S. Harding for his critical insights here.
11. Tyson Vaughan, interviews, Minamata, Japan, August 2010. I am grateful to Vaughan for allowing me to include this research story. The response of these individuals pushed me to consider more thoughtfully the political implications of envirotechnical analysis and inspired some of the discussion in this section. On Minamata, see George (2002) and Walker (2010). For parallels between Minamata and Fukushima, see Marran (2011).
12. I thank Julia Adeney Thomas for helping me contextualize Vaughan's research story.
13. However, on the contested definition of nuclearity, see Hecht (2010).

REFERENCES

Anderson, K. "A Hundred Days after Japan's Triple Disaster." *East Asia Forum,* June 20, 2011, accessed January 26, 2012. http://www.eastasiaforum.org/2011/06/20/a-hundred-days-after-japan-s-triple-disaster/.
Behr, P. "Desperate Attempts to Save 3 Fukushima Reactors from Meltdown." *New York Times,* March 14, 2011.
Biello, D. "Fukushima Meltdown Mitigation Aims to Prevent Radioactive Flood." *Scientific American,* June 24, 2011, accessed January 26, 2012. http://www.scientificamerican.com/article.cfm?id=fukushima-meltdown-radioactive-flood.
Bijker, W. "Globalization and Vulnerability: Challenges and Opportunities for SHOT around Its Fiftieth Anniversary." *Technology and Culture* 50 (2009): 600–612.
Bijker, W., T. Hughes, and T. Pinch (eds). *The Social Construction of Technological Systems*. Cambridge, MA: MIT Press, 1987.
Bocking, S. "Ecosystems, Ecologists, and the Atom: Environmental Research at Oak Ridge National Laboratory." *Journal of the History of Biology* 28 (1995): 1–47.
Boyd, W. "Making Meat: Science, Technology, and American Poultry Production." *Technology and Culture* 42 (2001): 631–664.
Bradsher, K., K. Belson, and M. Wald. "Executives May Have Lost Valuable Time at Damaged Nuclear Plant." *New York Times,* March 21, 2011.
Callon, M. "Society in the Making: The Study of Technology as a Tool for Sociological Analysis." In *Social Construction of Technological Systems,* edited by W. Bijker, T. Hughes and T. Pinch, 83–103. Cambridge, MA: MIT Press, 1987.
Clancey, G. *Earthquake Nation: The Cultural Politics of Japanese Seismicity, 1868–1930*. Berkeley: University of California Press, 2006.
CNN Wire Staff. "3 Japan nuclear reactors had full meltdown, agency says." *CNN.com,* June 6, 2011, accessed January 26, 2012. http://news.blogs.cnn.com/2011/06/06/3-japan-nuclear-reactors-had-full-meltdown-agency-says/.
Cronon, W. "The Trouble with Wilderness; or, Getting Back to the Wrong Nature." In *Uncommon Ground: Rethinking the Human Place in Nature,* edited by W. Cronon, 69–90. New York: Norton, 1995.
CTV.ca News Staff. "Tsunami Speed Comparable to Clip of Jumbo Jet." *CTV News,* March 11, 2011, accessed January 26, 2012. http://www.ctv.ca/CTVNews/TopStories/20110311/quake-feature-110311/.
Edgerton, D. *The Shock of the Old: Technology and Global History since 1900*. New York: Oxford University Press, 2007.
Edwards, P. Comments at "Infrastructure(s) and the Fukushima Earthquake: A Roundtable on Emergencies, Nuclear and Otherwise." Paper presented at the annual meeting of the Society for the History of Technology, Cleveland, OH, November 3–6, 2011.
Fountain, H. "A Look at the Mechanics of a Partial Meltdown." *New York Times,* March 13, 2011.
Gardner, R. "Constructing a Technological Forest: Nature, Culture, and Tree-Planting in the Nebraska Sand Hills." *Environmental History* 14 (2009): 275–297.
George, T. *Minamata: Pollution and the Struggle for Democracy in Postwar Japan*. Cambridge, MA: Harvard University Asia Center, 2002.
Geospatial Information Authority of Japan. "Land Subsidence Caused by 2011 Tōhoku Earthquake and Tsunami." April 14, 2011, accessed August 23, 2012. http://www.gsi.go.jp/sokuchikijun/sokuchikijun40003.html.
Gorman, H., and B. Mendelsohn. "Where Does Nature End and Culture Begin? Converging Themes in the History of Technology and Environmental History." In *The*

Illusory Boundary: Environment and Technology in History, edited by M. Reuss and S.H. Cutcliffe, 265–290. Charlottesville: University of Virginia Press, 2010.
Hamblin, J. *Poison in the Well: Radioactive Waste in the Oceans at the Dawn of the Nuclear Age*. New Brunswick: Rutgers University Press, 2008.
Hamblin, J. "Fukushima and the Motifs of Nuclear History." *Environmental History* 17 (2012): 285–299.
Haraway, D. *Simians, Cyborgs, and Women: The Reinvention of Nature*. New York: Routledge, 1991.
Haraway, D. *The Companion Species Manifesto: Dogs, People, and Significant Otherness*. Chicago: Prickly Paradigm Press, 2003.
Haraway, D. *When Species Meet*. Minneapolis: University of Minnesota Press, 2008.
Hecht, G. *The Radiance of France: Nuclear Power and National Identity after World War II*. 2nd ed. Cambridge, MA: MIT Press, 2009.
Hecht, G. "The Power of Nuclear Things." *Technology and Culture* 51 (2010): 1–30.
Hecht, G. "Nuclear Nomads: A Look at the Subcontracted Heroes." *Bulletin of the Atomic Scientists*, January 9, 2012.
Hilgartner, S. "The Social Construction of Risk Objects: Or, How to Pry Open Networks of Risk." In *Organizations, Uncertainties, and Risk*, edited by J.F. Short and L. Clark, 39–53. Boulder, CO: Westview Press, 1992.
Hughes, A., and T. Hughes. *Systems, Experts, and Computers: The Systems Approach in Management and Engineering, World War II and After*. Cambridge, MA: MIT Press, 2000.
Hughes, T. *Networks of Power: Electrification in Western Society, 1880–1930*. 3rd ed. Baltimore: Johns Hopkins University Press, 1993.
Hughes, T. *Human-Built World: How to Think about Technology and Culture*. Chicago: University of Chicago Press, 2004.
Jasanoff, S. (ed). *States of Knowledge: The Co-production of Science and the Social Order*. New York: Routledge, 2004.
Jones, C. "Defining the Problem." H-Energy, June 27, 2010, accessed January 26, 2012. http://www.h-net.org/˜energy/roundtables/Jones_Gulf.html.
Kelman, A. "Nature Bats Last: Some Recent Works on Technology and Urban Disaster." *Technology and Culture* 47 (2006): 391–402.
Kelman, A. "Boundary Issues: Clarifying New Orleans's Murky Edges." *Journal of American History* 94 (2007): 695–703.
Kline, R. *Consumers in the Country: Technology and Social Change in Rural America*. Baltimore: Johns Hopkins University Press, 2000.
Latour, B. *We Have Never Been Modern*. Translated by Catherine Porter. Cambridge, MA: Harvard University Press, 1993.
Latour, B. *Reassembling the Social: An Introduction to Actor-Network Theory*. New York: Oxford University Press, 2005.
Law, J. "Technology and Heterogeneous Engineering: The Case of Portuguese Expansion." In *Social Construction of Technological Systems*, edited by W. Bijker, T. Hughes and T. Pinch, 111–134. Cambridge, MA: MIT Press, 1987.
LeCain, T. *Mass Destruction: The Men and Giant Mines That Wired America and Scarred the Planet*. New Brunswick, NJ: Rutgers University Press, 2009.
Marran, C. "Contamination: From Minamata to Fukushima." *Asia-Pacific Journal* 9 (2011), accessed January 26, 2012. http://www.japanfocus.org/-Christine-Marran/3526.
Marx, L. "Technology: The Emergence of a Hazardous Concept." *Technology and Culture* 51 (2010): 561–577.
McCurry, J. "Fukushima Nuclear Plant Could Take 30 Years to Clean Up." *Guardian*, October 31, 2011, accessed January 26, 2012. http://www.guardian.co.uk/world/2011/oct/31/fukushima-nuclear-plant-30-years-cleanup.

McEvoy, A. "Working Environments: An Ecological Approach to Industrial Health and Safety." *Technology and Culture* 36 (1995): S145–172.
Mitchell, T. *Rule of Experts: Egypt, Techno-Politics, Modernity*. Berkeley: University of California Press, 2002.
Mitchell, T. "Carbon Democracy." *Economy & Society* 38 (2009): 399–432.
Mitchell, T. *Carbon Democracy: Political Power in the Age of Oil*. New York: Verso, 2011.
Monahan, A. "Tokyo Electric Power Delays Dumping Water at Fukushima Daiichi Plant." *Wall Street Journal,* April 11, 2011.
Murphy, M. *Sick Building Syndrome and the Problem of Uncertainty: Environmental Politics, Technoscience, and Women Workers*. Durham: Duke University Press, 2006.
Onishi, N., and M. Fackler. "In Nuclear Crisis, Crippling Mistrust." *New York Times,* June 12, 2011.
Parr, J. *Sensing Changes: Technologies, Environments, and the Everyday, 1953–2003*. Vancouver: University of British Columbia Press, 2010.
Pemberton, S. "Canine Technologies, Model Patients: The Historical Production of Hemophiliac Dogs in American Biomedicine." In *Industrializing Organisms: Introducing Evolutionary History,* edited by P. Scranton and S.R. Schrepfer, 191–213. New York: Routledge, 2004.
Perrow, C. *Normal Accidents: Living with High-Risk Technologies*. 2nd ed. Princeton, NJ: Princeton University Press, 1999.
Perrow, C. *The Next Catastrophe: Reducing Our Vulnerabilities to Natural, Industrial, and Terrorist Disasters*. Princeton, NJ: Princeton University Press, 2007.
Perrow, C. "Fukushima, Risk, and Probability: Expect the Unexpected." *Bulletin of the Atomic Scientists,* April 1, 2011, accessed November 4, 2012. http://thebulletin.org/web-edition/features/fukushima-risk-and-probability-expect-the-unexpected.
Pritchard, S. "Envirotech Methods: Looking Back, Looking Beyond?" Paper presented at the annual meeting of the Society for the History of Technology, Washington, D.C., October 17–21, 2007.
Pritchard, S. *Confluence: The Nature of Technology and the Remaking of the Rhône*. Cambridge, MA: Harvard University Press, 2011.
Pritchard, S. "An Envirotechnical Disaster: Nature, Technology, and Politics at Fukushima." *Environmental History* 17 (2012): 219–243.
Reuss, M., and S. Cutcliffe (eds). *The Illusory Boundary: Environment and Technology in History.* Charlottesville: University of Virginia Press, 2010.
Rosner, L. (ed). *The Technological Fix: How People Use Technology to Create and Solve Problems*. New York: Routledge, 2004.
Russell, E. "The Garden in the Machine: Toward an Evolutionary History of Technology." In *Industrializing Organisms: Introducing Evolutionary History,* edited by P. Scranton and S.R. Schrepfer, 1–16. New York: Routledge, 2004.
Sanger, D., and M. Wald. "Radioactive Releases in Japan Could Last Months, Experts Say." *New York Times,* March 13, 2011.
Schmid, S. "Both Better and Worse than Chernobyl." *London Review of Books,* March 17, 2011.
Shirouzu, N., P. Dvorak, Y. Hayashi, and A. Morse. "Bid to 'Protect Assets' Slowed Reactor Fight." *Wall Street Journal,* March 19, 2011.
Shulman, P. "A Catastrophic Accident of Normal Proportions." H-Energy, June 27, 2010, accessed January 26, 2012. http://www.h-net.org/~energy/roundtables/Shulman_Gulf.html.
Steinberg, T. *Acts of God: The Unnatural History of Natural Disaster in America*. New York: Oxford University Press, 2000.
Stine, J., and J. Tarr. "At the Intersection of Histories: Technology and the Environment." *Technology and Culture* 39 (1998): 601–640.

Sutter, P. "Comment on J. McNeill's *Mosquito Empires: Ecology and War in the Greater Caribbean,* 1620–1914." Paper presented at the annual meeting of the American Society for Environmental History, Tucson, AZ, April 13–15, 2011.

Tabuchi, H. "Company Believes 3 Reactors Melted Down in Japan." *New York Times,* May 24, 2011.

Tabuchi, H., and M. Wald. "Japanese Scramble to Avert Meltdowns as Nuclear Crisis Deepens after Quake." *New York Times,* March 12, 2011.

Tabuchi, H., and M. Wald. "Second Explosion at Reactor as Technicians Try to Contain Damage." *New York Times,* March 13, 2011.

Tabuchi, H., and M. Wald. "Japanese Reactor Damage Is Worse Than Expected." *New York Times,* May 12, 2011.

Tarr, J. "There Will Be Gas." *Pittsburgh Post-Gazette,* August 2, 2009, accessed January 26, 2012. http://www.post-gazette.com/pg/09214/987834-109.stm.

Thomas, J. *Reconfiguring Nature: Concepts of Nature in Japanese Political Ideology.* Berkeley: University of California Press, 2002.

US Geologic Survey. "Magnitude 9.0—Near the East Coast of Honshu, Japan." March 11, 2011, accessed January 26, 2012. http://earthquake.usgs.gov/earthquakes/recenteqsww/Quakes/usc0001xgp.php.

Walker, B. *Toxic Archipelago: A History of Industrial Disease in Japan.* Seattle: University of Washington Press, 2010.

Weiner, D. "A Death-Defying Attempt to Articulate a Coherent Definition of Environmental History." *Environmental History* 10 (2005): 404–420.

White, R. "Are You an Environmentalist or Do You Work for a Living? Work and Nature." In *Uncommon Ground,* edited by W. Cronon, 171–185. New York: Norton, 1995.

World Health Organization. "The Great East Japan Earthquake." Geneva: WHO Press, 2012, accessed August 23, 2012. http://www.wpro.who.int/publications/docs/japan_earthquake.pdf.

Yomiuri Shimbun. "Nuclear Crisis: How It Happened, Safety Vows Forgotten, 'Safety Myth' Created." *Yomiuri Shimbun,* June 15, 2011.

8 Nuclear Power after 3/11

Looking Back and Thinking Ahead

Catherine Butler, Karen A. Parkhill, and Nicholas F. Pidgeon

Within science, technology and society studies (STS), nuclear energy has for many years formed a key focus of analyses and empirical study. Indeed, with regard to the branch of STS concerned with the relationships between risk, science, technologies, and publics, nuclear energy can perhaps be seen as a paradigm case (e.g. see Beck 1992, Welsh 2000, Wynne 1992). Studies have examined a wide range of issues, resulting in a well-established social science basis for interrogating and understanding various dimensions of nuclear energy. This includes analyses focused on the framing of nuclear safety and the governance of risk (e.g. Jasanoff 2005, Meshkati 1991, Renn 2008). It encompasses research concerned with the significance of public engagement, opinion, knowledge, and local communities in the context of nuclear energy development (e.g. Butler et al. 2011a, Eiser et al. 1995, Parkhill et al. 2010, Pidgeon et al. 2008a, Wynne 1992). There are projects directed at understanding the ethical implications of nuclear power, including those that draw from environmental justice literature as their basis for departure (e.g. Blowers and Leroy 1994, Endres 2009, Stanley 2009). We also see works dedicated to analysis of the media framing of nuclear energy and its implications for wider public and political engagement (e.g. Doyle 2011, Gamson and Modigliani 1989). This highlights just a few of the themes long established in the social scientific study of nuclear energy.

Following the declaration of nuclear emergency in Japan, questions surrounding the use of nuclear power have reappeared at the forefront of public debate. The 9.0 magnitude earthquake and following tsunami that struck Japan on March 11, 2011, had devastating consequences for many people. The subsequent problems encountered at Japan's nuclear power plants, and particularly at Fukushima Daiichi, have raised questions about the future of nuclear energy worldwide. In response to the earthquake, the six nuclear reactors at the Fukushima Daiichi site were all safely shut down but subsequent power outages caused by the tsunami resulted in a failure of the cooling systems, eventually leading to a release of radioactive material across four units (IAEA 2011, also see Hale 2011). The accident was rated as a level 7 'major accident' on the International Nuclear and Radiological Event Scale (INES), used to describe an event involving a 'major release of radioactive material with widespread health and environmental effects requiring implementation of planned and extended countermeasures' (IAEA 2011).

These events came at a time of projected global resurgence in nuclear energy facility development, with an estimated 360 gigawatts of additional nuclear generating capacity projected to be developed worldwide by 2035, on top of the 390 gigawatts already in use (IEA 2010). This renewed interest in nuclear was, in part, due to its potential as a low carbon energy source but was also related to concerns about energy security as energy demand and competition for hydrocarbon fuels increase worldwide. In the context of this so-called nuclear renaissance, the technology has, however, remained contentious, provoking vociferous public debate and, in the UK at least, even legal battles (see Greenpeace 2007). The full extent as to what the events in Japan might mean for nuclear power in this contemporary context of renewed political and market interest is yet to be seen. However, the long standing history of studying nuclear energy within the STS tradition, as well as within social science more broadly, provides a strong basis for thinking through some of the implications for nuclear energy in light of this accident. In this chapter, we reexamine existing social science research on nuclear energy to think through some of the key sociopolitical dimensions of concern with regard to nuclear power after Fukushima.

We do not, of course, in this brief treatment address all existing social science aspects pertinent to this issue. Instead, we aim to cover key aspects, structuring our discussion around four sections themed according to major areas of research that have addressed nuclear energy as a substantive issue. The themes are clustered as follows: policy, political acceptability, and economics; public opinion and attitudes; safety, justice, and ethics; and framing and the media. Through these interlinked sections we aim to provide an overview of some of the central issues and findings in STS and social science research on nuclear energy and (re)contextualize key ideas with reference to the still unfolding disaster in Japan.

POLICY, POLITICAL ACCEPTABILITY, AND ECONOMICS

Policy and the political acceptability of nuclear energy along with the associated economics have been the focus of much academic commentary and debate. Beck (1992, 1999), for example, has detailed how political acceptability has been achieved through a conflation of risk and control—'the greater the risk, the greater the need for controllability' (1999: 6). For Beck, this has allowed the limits of controllability to be brushed from view and facilitated the political acceptance of hazards—transposed to risks that can be managed (also see van de Poel 2011). In a different but closely related line of enquiry, Jasanoff (2005) has pointed to the concepts of *risk* and *precaution* and their uptake within different national policy contexts as important in defining the manifestations of environmental policy. She too points to the way that risk framings have facilitated political acceptance of risks deemed to be small (even if their consequences are high). These analyses of the different logics that inform policy and provide openings for the pursuit of particular technologies or approaches can be useful in unpacking the unfolding policy responses to the Fukushima accident in different national contexts.

At the time of the Fukushima Daiichi accident the political acceptability of new nuclear energy had been reinvigorated with the emergence of a framing of nuclear as a solution to policy aims to decarbonize electricity and address energy security (Bickerstaff et al. 2008, Duffy 2011). Even for countries, such as Germany, with no plans to pursue *new* nuclear power, life extensions to existing plants were seen as an essential 'bridging technology' for transitioning to low-carbon energy systems (Federal Ministry of Economics and Technology 2010). Though the full implications of the events at Fukushima continue to ripple through global energy policy discourses, it is possible to identify two broad policy pathways adopted following the accident: (1) amplification of risk and withdrawal of policy support, and (2) safety review, then attenuation (reduction) of risk, followed by continued support (on amplification and attenuation see Pidgeon et al. 2003).

Prominent examples of the first have been unfolding in Japan, Germany, Switzerland, and Italy. It is perhaps unsurprising that Japan withdrew support due to a reconceptualization of the scale of the risks involved, with former prime minister Naoto Kan (2011) specifically giving the example of the evacuation zone as an impetus. Potentially more startling has been the reaction of the German government, which, due to the 'residual risks' of nuclear power, has rescinded support for operating-life extensions and instead aims to completely phase out nuclear power plants by 2022 (Federal Ministry for the Environment, Nature Conservation and Nuclear Safety 2011). In contrast, the UK and US have situated the events at Fukushima as part of 'learning from experience'. This has allowed for continued adherence to the 'principle of *continuous improvement*', meaning further new nuclear power occurring at the same time with the development of more safety measures, procedures, and knowledge (HM Chief Inspector of Nuclear Installations 2011). In this, the UK has taken a lead, solidifying its support for eight new nuclear plants through the parliamentary approval of its Energy National Policy Statements (DECC 2011).

This suggests that Fukushima has ratcheted up current policy trends and logics in some countries, rather than stimulating a wholesale political reversal. We can understand these differing trends through reference to the varying emphasis placed on prevailing logics of *risk* or *precaution*. In the US and UK the favored term has been that of risk, while in Germany, where the notion of the precautionary principle originates, we can see a general propensity toward precaution (Jasanoff 2005). These different terms embed different ways of thinking about and approaching policy that can be seen as underlying factors in the emergence of very different responses to the Fukushima disasters across different national policy contexts.

In terms of economics, the events in Japan serve to refocus attention on questions concerning the evolving relationship between private financing of nuclear energy, sociopolitical acceptability, and nuclear accidents. In the UK, development of new nuclear generating facilities is agreed on the basis that development is privately financed without government subsidy (Hendry 2011), while the US administration has taken the route of offering conditional loan guarantees (The White House 2011). The assertions governments around the world have made regarding the role of public investment in nuclear are, however, complicated by the clear implications of the accident for government spending in Japan. The expenditure for compensation alone is estimated to be US$124 billion, the costs of which are being covered in the first instance by special government-issued bonds that the private owners of the Fukushima power station (Tokyo Electric Power Company, or TEPCO) will be expected to repay over an as yet unspecified number of years (McCurry 2011). Negotiations over this repayment policy occur at the same time as TEPCO's share price plummets, resulting in concerns that the costs will be passed onto customers, leading to a double burden for the wider public (i.e. in paying through both taxes and bills). In this context, notions of a completely privately financed nuclear sector are problematic, as in the event of an accident of any magnitude the role of government subsidies and loans/financial guarantees becomes significantly complicated.

The global trends in private financing and nuclear energy have been the subject of scrutiny both prior to and after the accident at Fukushima Daiichi. Even before Fukushima, investor decisions had been affected by 'reputational' risks associated with nuclear energy, for example, the withdrawal of Deutsche Bank and HypoVereinsbank from the financing of the Belene nuclear power station (Bulgaria) following protests by an internationally networked civic group (Beck 2009). After the accident, such reputational risks are heightened further and combined with closely linked financial risks to raise questions about nuclear investment. This has led to speculation regarding whether these events might lead to greater investment in renewable forms of energy or surges of investment in hydrocarbons.

In their 'Lower Nuclear' scenario, the International Energy Agency (IEA) envisions that in the event of lower than anticipated nuclear development, increased investment in hydrocarbons would lead to a 30% higher growth in emissions than in their 'New Policies' scenarios (IEA 2011). Such scenarios can, however, be critiqued for being overly optimistic in their projections of global increases in nuclear generating capacity as well as lifetime extensions, thus overestimating the contribution to global emissions reductions that nuclear would make (Jewell 2011, Schneider et al. 2011). The potential for reduced investment in nuclear energy has resulted in further concerns about possible increased competition over hydrocarbons and the accompanying energy security issues. This is particularly relevant in countries where indigenous resources are limited and where policies to phase out nuclear power have been adopted.

For the time being, it remains unclear as to how far investment in nuclear energy will be affected by the Fukushima accident in the long term and, indeed, what the implications of this will be for the global energy market, carbon emissions, or energy security. There are some indications of the impact, however, evident in the difficulties the UK is facing in securing financial investment for its new nuclear build program (e.g. see Gosden 2012). The extent to which nuclear energy development would have ever reached the levels some were projecting had Fukushima not happened will of course always be open to question.

PUBLIC OPINION AND ATTITUDES

Nuclear power has—beyond its beginnings where 'glamorous reactors' were anticipated with 'a great sense of excitement'—had a tumultuous relationship with the public (Welsh 2000: 1). It has been characterized as a 'uniquely dreaded' technology due to its long-standing association with atomic weaponry, invisible and long-lasting effects of radiation, and concerns about waste disposal (Masco 2006, Slovic 1997). In the 1980s, after the nuclear incident at Three Mile Island (1979) and the disaster at Chernobyl (1986), public opposition to nuclear power was at an all-time high in many countries. Indeed, data from the US even before Chernobyl suggested that public opposition to nuclear new build rose from around 20% in the 1970s to more than 60% in the early 1980s (Rosa and Freudenburg 1993). Other research has identified public distrust of regulators, government and the nuclear industry to manage risks responsibly and provide truthful information to the public as a key reason for erosion of support (Wynne 1992, 1996).

In the period 2000–2010, opinion polling indicated a reduction in opposition. For example, a global poll by the Organization for Economic Cooperation and Development (OECD) and the Nuclear Energy Agency showed in 2010 that support for nuclear energy had increased in countries such as the US, Japan, Sweden, Finland, and the UK (OECD 2010). Looking specifically at the UK, polling of the British public conducted in early 2010 found a very balanced picture, with 46% of those questioned favoring replacement or expansion of the existing nuclear capacity in Britain as compared to 47% who wanted it closed or phased out at the end of the existing program (Spence et al. 2010). However, a closer look at the national polling data shows a more complex picture, with a large proportion of recent national support remaining fragile in having a conditional or 'reluctant acceptance' at best (Bickerstaff et al. 2008, Corner et al. 2011, Pidgeon et al. 2008b).

Shortly after the incident at Fukushima Daiichi we posited that a short- to medium-term impact of the disaster would be that many might withdraw their support for nuclear power and in particular for nuclear new build (Butler et al. 2011b). Thus, opposition during that period would correspondingly increase. Early polling research suggested this was exactly the case, with a rise in opposition that outweighed support identifiable in many countries, though in some cases only by the thinnest of margins. The US is a notable exception where support for nuclear power remained marginally higher than opposition (see Figure 8.1) (Ipsos MORI 2011).

In the same poll, more than half of the respondents from Japan who indicated they now opposed nuclear energy to produce electricity said they did so due to the events in their country; significant proportions of the public in other countries also stated that was the case (see Figure 8.2) (Ipsos MORI 2011). On the basis of such findings, we might expect that those communities who are proposed as hosts for a new reactor may now oppose such developments.

For communities with no experience of a nuclear facility, it is likely that within the short to medium term, potential public contestation surrounding nuclear power may indeed prove to be a stumbling block. However, this is not necessarily true of all proposed reactor sites. For example, in the UK proposed sites are either on or adjacent to an existing nuclear power station. Previous research tells us that the response of people in such communities does not always mirror that obtained from national samples. While reluctant acceptance may be a feature of discourse in such communities and Fukushima may prompt the 'extraordinariness' of living close to a nuclear facility to cause momentary reframings of nuclear power as a threat, there

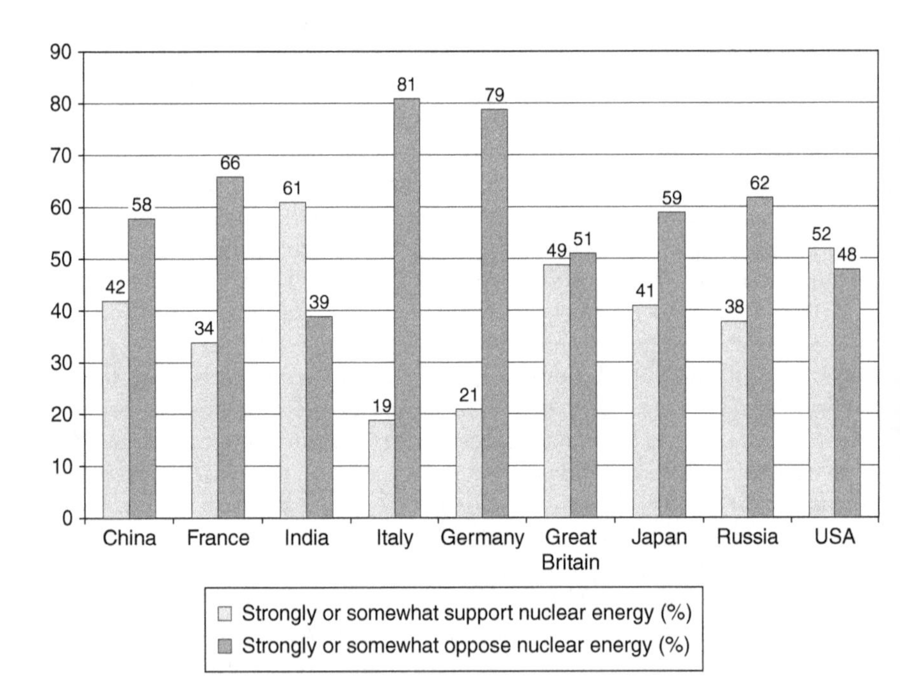

Figure 8.1 '[Nuclear energy] Please indicate whether you strongly support, somewhat support, somewhat oppose, or strongly oppose each way of producing electricity', poll conducted between April 6 and 21, 2011.

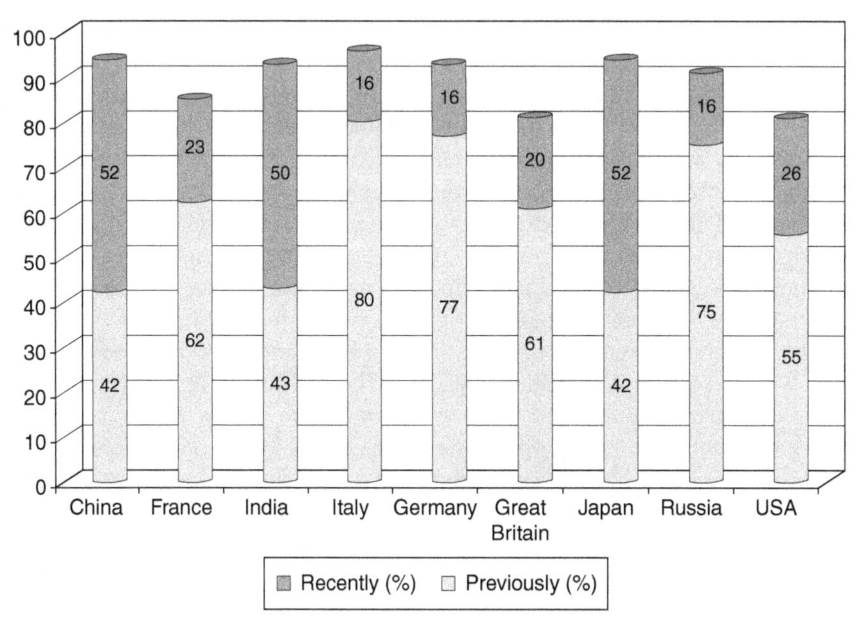

Figure 8.2 'You indicated that you oppose nuclear energy to produce electricity. Have you held this view previously or have you decided recently to oppose it because of events in Japan?', poll conducted between April 6and 21, 2011.

are some important qualitative nuances to public perceptions that may lead to differing medium to long term trends following Fukushima (Henwood et al. 2010, Parkhill et al. 2010, Pidgeon et al. 2008a).

Examples include the importance of social familiarity, which through social networks connected to the power station (i.e. either being or knowing a power station worker) or through imaginary positioning (being able to *imagine* how workers think, feel, and follow working practices) demystifies the power station as a distant institutional organization (Parkhill et al. 2010). As such, trust in power station workers is engendered. Although hidden anxieties may come to the surface in light of Fukushima, these could also be moderated by the distancing of the events as irrelevant to localized contexts and working practices, serving to reify the perceived safety of local plants (and trust in plant operators) rather than undermining it (Parkhill et al. 2011, Zonabend, 1993).

SAFETY, JUSTICE, AND ETHICS

The accident at Fukushima Daiichi brings to the forefront of debate the justice and ethical issues surrounding the safety of nuclear power and the consequences when something goes wrong (e.g. contamination, environmental degradation, displacement, and health). As noted at the beginning of this chapter, both the STS tradition and the field of environmental justice have made significant contributions to the examination of safety, justice and ethical concerns with regard to nuclear energy (e.g. see Meshkati 1991, Shrader-Frechette 2002, Wynne 1992). Existing ideas within social science thus have a great deal to offer in addressing questions at the intersection of safety, ethics and justice in the aftermath of the disaster. Indeed, a significant part of the post-Fukushima social science analysis has addressed itself to questions around safety measures, highlighting the significance of human factors in safety assessments and the connections to ethical concerns (e.g. see Moller and Wikman-Svahn 2011, Shockley 2011, Shrader-Frechette, 2011a).

Fukushima has prompted many governments to review nuclear energy risk governance and safety procedures. Inspections in the US, which examined the capacity of the 104 operating US nuclear plants to deal with power losses or damage to large areas of a reactor site following extreme events, concluded that 'all the reactors would be kept safe even in the event their regular safety systems were affected by these events, although a few plants have to do a better job maintaining the necessary resources and procedures' (US Nuclear Regulatory Commission 2011). Similarly, the UK Office for Nuclear Regulation initiated a review, focusing primarily on the safety of the existing gas-cooled reactor fleet in the UK (which it argues are inherently more resilient than light water reactors to loss of power and other external shocks). Their resultant interim report concluded: 'In considering the direct causes of the Fukushima accident we see no reason for curtailing the operation of nuclear power plants or other nuclear facilities in the UK' (Office for Nuclear Regulation 2011). These conclusions stand in stark contrast to the German Ethics Commission for a Safe Energy Supply report (2010), which states: 'The withdrawal from nuclear energy is necessary and is recommended to rule out future risks that arise from nuclear in Germany. It is possible because there are less risky alternatives'. The German report adopts a wider framing than the UK analysis, arguing (1) that if such a technologically advanced nation as Japan could suffer this catastrophe, this undermines the assumption made in Germany that a major nuclear accident could not happen there, and (2) that the considerable remaining uncertainties surrounding the temporal, financial, and geographical scope of the accident also undermine the belief that a nuclear accident can be sufficiently contained in heavily populated countries such as Germany or Japan.

These very different approaches can be viewed as related to differences in emphasis given to concepts of risk or precaution, discussed earlier, but can also be seen as tied to the way Fukushima as an event was subsequently positioned in relation to these different underlying logics. Moller and Wikman-Svahn (2011) draw a distinction between black swan events—ones only predictable in hindsight—and black elephant events—ones that are predictable but ignored (partially because they are deemed improbable).

The German position reveals a shift that appears to both reconstitute nuclear accidents as having been black elephants, and place greater emphasis on the associated uncertainties. Whereas, the UK and US positions reveal a tendency toward classifying Fukushima as a black swan event, leading to the conclusion that existing nuclear designs and reactors are safe as accidents are still considered to be low probability events that therefore do not warrant increased concern or changes in policy.

In the context of the German government's decision, justice and ethical issues associated with nuclear energy are more explicitly discussed and play a far greater role in the reasoning about safety in a post-Fukushima context. While the German Ethics Commission for a Safe Energy Supply places such concerns at the very center of thought about safety, in other political contexts, such as the UK, ethical issues are posed as a trade-off with those that arise as a consequence of climate change and energy security (e.g. see Department of Trade and Industry 2007: 25). Nuclear energy is positioned as an antidote to the current and future justice and ethical issues related to climate change and energy security, and the injustices associated with nuclear energy are to be endured as part of the cost of the solution. In this context, the approach in Germany has come under attack for not prioritizing the ethical issues that arise with climate change over those that arise with nuclear energy (e.g. see Lynas 2011a). However, in the German case, the withdrawal from nuclear energy is premised on the notion that there are less risky alternatives via which these issues can be addressed (Ethics Commission for Safe Energy Supply 2011). In this respect, the necessity of nuclear energy for meeting climate change targets and delivering a secure energy system has been the subject of vociferous debate. Although, as discussed earlier, there are scenarios that envision significant increase in global emissions and energy insecurity as a consequence of decisions not to pursue nuclear, these projections have been heavily critiqued (e.g. see Schneider et al. 2011, Shrader-Frechette 2011b).

Climate and energy justice are complex areas in their own right and refer not only to issues associated with the impacts of climate change (e.g. droughts, floods, displacement, and so forth) and energy security but also to the burdens and responsibilities associated with mitigation and adaptation efforts and energy systems more widely (Klinsky and Dowlatabadi 2009). Thus, ethical and justice issues posed by climate change and energy are far more complex and wider ranging than a simple question of reducing emissions and securing supplies versus consequences of impacts. The characterization of the justice issues as simply a trade-off between climate change and energy security, on the one hand, and nuclear energy risks, on the other, can therefore be seen as problematic. This simplistic portrayal fails to engage properly with the climate and energy justice issues (and what the range of solutions to those might be), and serves to redirect attention away from the multiple questions and issues to be addressed regarding the justice and ethical concerns that nuclear energy itself raises (e.g. see Blowers 2011, also in this book).

Such issues range from the displacement of communities and loss of livelihoods, to the exposure of workers to high levels of risk (Shrader-Frechette 2002). This latter concern is particularly salient for the Japanese case, as the specification of legal levels of radiation was modified early in the crisis to increase the annual radiation exposure limit from 100 mSv (millisievert) a year to 250 mSv a year to allow workers to spend more time at the plant where nuclear fuel in three reactors suffered meltdowns (McCurry 2011, also see Kermisch 2011, Pritchard in this book: 124). The malleability of such levels, questions about accurate measurement, and the implications of exposure for workers and their families are undoubtedly areas that should be the subject of further scrutiny in building understanding of nuclear accidents.

In her analysis of post-Chernobyl Ukraine, Petryna (2002) explores the way that citizenship came to encompass rights to the acknowledgment and compensation of biological injury. In light of this, it is interesting to reflect on the unfolding compensation debate in Japan, which has been commanded as much by financing concerns as by the citizenship rights of those affected by the accident. Initial discussions involved proposals for the capping of compensation so as to ensure TEPCO would not be made bankrupt by compensation claims. But these plans were subsequently supplanted by moves made by the Japanese government to underwrite compensation, providing TEPCO did not cap compensation payouts. In this context, important questions around the extent to which financial compensation is ever an effective or viable ethical response to biological injury caused by exposure to technological risks have been left largely unanswered (Hoffman 2001).

FRAMING AND THE MEDIA

Previous research has identified that the role of news media as a purveyor of interpretive packages is of high importance in the case of policy issues, particularly controversial ones like nuclear energy (Gamson and Modigliani 1989). The media plays a significant part in the relationships between government and public(s) and in developing and crystallizing the cultural meanings that circulate. This interaction involves interplay between media interpretative packages and 'the life histories, social interactions and psychological predispositions', which people bring to the process of constructing meaning (Gamson and Modigliani 1989: 2). There is, then, no direct or linear connection between public opinion and changes in media discourse but the wider role media plays in communication as one part of the process through which public(s) construct meanings is nevertheless significant. People are thus active in making sense of the world but this requires work and, as such, the frames that are developed, spotlighted and readily accessible have a higher chance of being used.

The interpretive packages that news media provide 'have the task of constructing meaning over time, incorporating new events into their interpretive frame' (Gamson and Modigliani 1989: 10). In the nuclear case, the ability of interpretive packages to incorporate events like Fukushima is paramount to the plausibility and consistency of those frames. A number of existing interpretative frames or packages for nuclear energy have been identified in previous research (Gamson and Modigliani 1989). First is the *progress* frame; this entails the identification of nuclear fission as holding immense potential, either for good (power generation and production) or for bad (weaponry and destruction). Second is the *energy independence* discourse, initially identified as emerging in media rhetoric during the 1970s oil crisis but also strongly represented in contemporary debates (see Corner et al. 2011).

Third, and counter to these, are three further frames, all of which correspond with an antinuclear narrative: *soft paths, public accountability*, and *not cost-effective*. The *soft paths* interpretative package entails a critique of nuclear power based on the kind of path on which it places our societies, that is, one that entails a culture wasteful of energy, involving highly centralized technologies, and insensitive to the ecological consequences. The *public accountability* narrative emphasizes an anticorporate message, and the *not cost-effective* frame takes up the economic questions around nuclear energy. Fourth, the *runaway* narrative is identified, which while retaining an antinuclear flavor is more resigned than opposed, more of a 'grin and bear it' narrative than a 'no nukes' message (Gamson and Modigliani 1989: 20). Finally, a *devil's bargain* discourse is evident, which is a combination of the pronuclear interpretive packages (*progress* and *energy independence*) and the *runaway* frame. In the more recent discursive context, we might view the framing of nuclear as a low-carbon energy source in the context of climate change as part of this *devil's bargain* narrative (see Bickerstaff et al. 2008, Doyle 2011, Pidgeon 2012).

The Fukushima Daiichi accident can be seen as having provoked a 'critical discourse moment', wherein the culture, frames, and interpretative packages of an issue become visible (Chilton 1987, Gamson and Modigliani 1989). The early media reporting and wider public discourse surrounding the events in Japan revealed tendencies toward the positioning of discussion across the broad interpretative frames discussed above. Rather than raising new debates and issues, the accident has acted as a 'peg' for preexisting interpretive packages and has been used in support for *or* critique of existing cultural viewpoints relating to nuclear energy. In the Japan case we find numerous examples where existing interpretative packages have been adapted to the events (see Table 8.1).

These broad frames circulate through media and wider public discourse, reemerging as events unfold to provide interpretations that fit with wider cultural ideas. Their importance is in part found in what they reveal about the way different cultures relate to nuclear energy. A more extensive examination of media content following the accident at Fukushima could reveal further emergent interpretative frames. Moreover, study of the levels at

Table 8.1 Illustrative examples of interpretive packages in post-Fukushima news media coverage

Interpretive packages/ frames	Example extracts from news media coverage
Progress	"Why Fukushima made me stop worrying and love nuclear power . . . A crappy old plant with inadequate safety features was hit by a monster earthquake and a vast tsunami. The electricity supply failed, knocking out the cooling system. The reactors began to explode and melt down. The disaster exposed a familiar legacy of poor design and corner-cutting. Yet, as far as we know, no one has yet received a lethal dose of radiation." (Mombiot, 2011)
Energy independence	"The recent disaster in Fukushima has set public confidence in nuclear power back to levels not seen since the aftermath of the Chernobyl or Three Mile Island disasters. This really is a shame, because I believe that nuclear power, if the proper precautions are taken, could greatly lessen the current dependency for fossil fuels, something which is direly needed." (Boisvert, 2011) "The Germans topped that by switching off several nuclear power stations unnecessarily and importing millions more tonnes of coal (the biggest killer of all energy sources by some margin) from the United States to keep the lights on." (Lynas, 2011a)
Soft paths	"100% renewables (and geothermal) is where we need to get to eventually—so why not seek to get there just as soon as possible without yet another disastrous foray into today's nuclear cul-de-sac?" (Porritt, 2011)
Public accountability	"Investigators may take months or years to decide to what extent safety problems or weak regulation contributed to the disaster at Daiichi, the worst of its kind since Chernobyl. But as troubles at the plant and fears over radiation continue to rattle the nation, the Japanese are increasingly raising the possibility that a culture of complicity made the plant especially vulnerable to the natural disaster that struck the country on March 11. . . . The mild punishment meted out for past safety infractions has reinforced the belief that nuclear power's main players are more interested in protecting their interests than increasing safety." (Normimitsu and Belson, 2011)
Not cost-effective	"Fukushima shows us the real cost of nuclear power . . . The economics of nuclear power don't add up—which is even more reason to invest in renewable energy." (Bennett, 2011)

Table 8.1 (Continued)

Interpretive packages/ frames	Example extracts from news media coverage
Runaway	"The twin natural disasters have also turned the Fukushima Daiichi nuclear power plant into Frankenstein's monster, a man-made object threatening man." (Rani, 2011)
Devil's bargain	"There is no doubt that the explosions and radioactive releases at the stricken Fukushima Daiichi plant represent the worst nuclear disaster since the explosion at the Chernobyl power plant in Ukraine in 1986. However . . . if we abandon nuclear, prepare for a future of catastrophic global warming, imperilling the survival of civilisation and much of the earth's biosphere." (Lynas, 2011b)

which different interpretative packages received representation in the media following Fukushima could also be of significant interest.

The media serve to reflect, create, and crystallize cultural meanings; this communicating role is particularly salient in the context of controversial science and technological issues, with nuclear power being a paradigm case. Several core lessons have emerged from what has been termed 'risk communication' research on which we can draw to reflect on the media and wider public discourse surrounding the Fukushima accident. These include reference to the importance of dialogue rather than one-way communication: enabling trust, exploring divergent values of varied public(s); meeting concerns about governance arrangements; and not treating publics as irrational but rather recognizing that their responses to risk may be rooted in different concerns (Fischhoff 1995, Pidgeon and Fischhoff 2011). Many of the public and media statements about nuclear risk following Fukushima appear to have failed to take into account these research insights (see also Friedman 2011). In terms of political and public acceptability, the conflicts that exist between the media being a key player in the construction of meaning around nuclear issues, and its incompatibility with the requirements of complex risk issues in the delivery of messages, may serve to heighten difficulties in establishing meaningful debate in an already charged and difficult area.

CONCLUSIONS: REFLECTIONS ON NUCLEAR POWER AFTER FUKUSHIMA DAIICHI

In concluding, we want to draw together some of the interlinkages and key arguments that have emerged through our analysis. The role of the media in framing, (re)circulating, and solidifying cultural meanings has clear interconnections with the development of public opinion. Analysis of media can offer some insight into the cultural meanings that circulate with regard to nuclear and help to build a picture of the themes flowing through wider public discourse. In turn, research into public attitudes and opinions provides insight into, at a more general level, the sway of public feeling on an issue and, more specifically, informs us about cultural frames that do not appear in media discourse (e.g. familiarity frames) (Parkhill et al. 2010). Both media representation and public opinion are also interwoven in multiple ways with political acceptability and economic issues. Political actors can been seen both as 'sponsors', with roles in narrating particular storylines or developing interpretative packages (Gamson and Modigliani 1989: 6), and as end users of media to gain insights into the cultural frames that are circulating in wider public discourse. In addition, media representations play a significant role in influencing views regarding economic investment and financial risk, particularly when we consider the role of reputational risks in financing decisions (see Beck 2009).

It seems fair to argue that while justice issues are generally regarded as important, in most national contexts they have not taken a front seat in decisions or in general debate around nuclear energy. This can have strong implications for decisions in this area. As Stirling (1997) points out, if intergenerational equity as one of a number of criteria (e.g. the irreversibility of the technology, the imposed nature of risks) weighed heavily in decision making, this would overturn the advantage that nuclear power has over other energy options. How we draw boundaries and what is weighted most heavily in debates can drastically affect decisions and outcomes. In the face of this, Stirling concludes that the importance of procedural justice in making energy policy decisions may be heightened: 'Ultimately, the only satisfactory way to address issues of divergent value judgements is through political discourse and democratic accountability' (Stirling 1997: 535).

This suggests, if the aim is for a more 'just' energy system, procedures that open up discourse—for example, beyond a limited number of interpretive frames and with a wider boundary that encompasses other aspects of the energy system beyond nuclear—will be important for advancing decision making. This argument will have been significant before Fukushima but the accident serves to provide a window to open up debate about these issues and more. Though the accident itself may not have provoked new cultural discourses around nuclear power, instead acting as a catalyst for the reemergence of existing frames and debate, the potential for reflection and emergent themes is now present.

The accident brings into focus both the fragility of nuclear energy and its clear durability. These events demonstrate the strong capacity of the sector to withstand significant opposition and maintain political favor in such a context. Though discourses of progress, energy independence, and now climate change may increasingly have to compete with multiple critical framings, they represent strong recurring themes that hint at the fundamental staying power of nuclear. However, Fukushima also provides an indication of the fragility of nuclear power in terms of the potential such an accident can have for triggering distrust toward the safety of nuclear and toward those institutions responsible for such issues, with serious implications for investment, political acceptability, and future development. These aspects of fragility have the potential to intensify if, for example, any hint of a 'cover-up' emerges or even a lack of transparency is perceived. Implications of this kind are evident in the Fukushima case, with three Japanese officials fired in the early aftermath of the accident in an attempt to rebuild public trust in the relations between government and industry (Mogi 2011a).

We might see the dramatic change in Japan's energy policy and the 'ramping up' of existing concerns and political positions in countries like Germany and Italy as indicative of this fragility. However, we also begin to see evidence of the durability of nuclear power as these countries come under pressure to find ways of meeting their energy demand and address climate change targets (Fugino 2011, Mogi 2011b). The wider implications of Fukushima might thus be found in what arises from the efforts in both Germany and Japan to decarbonize and ensure the security of their energy systems without the use of nuclear power in the long term. Such a shift away from nuclear energy by some nation states is likely to have far-reaching implications for the prospect of any 'nuclear renaissance' globally. Already in the UK we have seen a number of consortia withdraw their applications for the previously coveted new build sites. In many senses then, the outcomes of the approaches to contemporary energy policy in countries like Germany may have far greater implications for the future of nuclear energy than can be seen in the immediate sociopolitical reactions to the disaster itself. In particular, implications for the development of renewable energy technologies and carbon capture and storage could be highly significant as efforts to create low carbon systems without nuclear energy unfold. Ultimately, if they are successful in creating low-carbon, secure energy systems without nuclear this could alter the trajectory of energy system development more widely around the world.

ACKNOWLEDGMENTS

The authors would like to acknowledge the support of the Leverhulme Trust (F/00 407/AG), the UK Energy Research Centre (NE/G007748/1), and the Interdisciplinary Cluster on Energy Systems, Equity and Vulnerability (InCluESEV).

REFERENCES

Beck, U. *Risk Society: Towards a New Modernity*. London: Routledge, 1992.
Beck, U. *World at Risk*. Cambridge: Polity Press, 1999.

Beck, U. *World at Risk*. Cambridge: Polity Press, 2009.

Bennett, C. "Fukushima Shows Us the Real Cost of Nuclear Power." *Guardian Online*, accessed March 21, 2011. http://www.guardian.co.uk/commentisfree/2011/mar/23/fukushima-nuclear-power-renewable-energy.

Bickerstaff, K., I. Lorenzoni, N. Pidgeon, W. Poortinga, and P. Simmons. "Reframing Nuclear Power in the UK Energy Debate: Nuclear Power, Climate Change Mitigation and Radioactive Waste." *Public Understanding of Science* 17 (2008): 145–169.

Blowers, A. "Why Fukushima Is a Moral Issue: The Need for an Ethic of the Future in the Debate about the Future of Nuclear Energy." *Journal of Integrative Environmental Sciences* 8, no. 2 (2011): 73–80.

Blowers, A., and P. Leroy. "Power, Politics and Environmental Inequality: A Theoretical and Empirical Analysis of the Process of 'Peripheralisation.'" *Environmental Politics* 3 (1994): 197–228.

Boisvert, M. "Fukushima: Setting Energy Independence Back Again." *BardPolitik-Daily,* accessed April 17, 2011. http://bardpolitikdaily.blogspot.com/2011/04/fukushima-setting-energy-independence.html.

Butler, C., K. Parkhill, and N. Pidgeon. "Nuclear Power after Japan: The Social Dimensions." *Environment: Science and Policy for Sustainable Development,* 53, no. 6 (2011a): 3–14.

Butler, C., K. Parkhill, and N. Pidgeon. "From the Material to the Imagined: Public Engagement With Low Carbon Technologies in a Nuclear Community." In *Renewable Energy and The Public: From NIMBY to Participation*, edited by P. Devine-Wright, 301–315. London: Earthscan, 2011b.

Chilton, P. "Metaphor Euphemism and the Militarisation of Language." *Current Research on Peace and Violence* 10, no. 7 (1987): 7–19, cited in A. Gamson and A. Modigliani, "Media Discourse and Public Opinion on Nuclear Power: A Constructionist Approach," *American Journal of Sociology* 95, no. 1 (1989): 1–37.

Corner, A., D. Venables, A. Spence, W. Poortinga, C. Demski, and N. Pidgeon. "Nuclear Power, Climate Change and Energy Security: Exploring British Public Attitudes." *Energy Policy* 39 (2011): 4823–4833.

DECC (Department for Energy and Climate Change). "National Policy Statement for Nuclear Power Generation Volume 1 (EN-6)." London: Crown Copyright, 2011.

Department of Trade and Industry (DTI). *The Future of Nuclear Power: The Role of Nuclear Power in a Low Carbon UK Economy*. London: Crown Copyright, 2007.

Doyle, J. "Acclimatizing Nuclear? Climate Change, Nuclear Power and the Reframing of Risk in the UK News Media." *International Communication Gazette* 73, nos. 1–2 (2011): 107–125.

Duffy, R. "Déjà Vu All over Again: Climate Change and the Prospects for a Nuclear Power Renaissance." *Environmental Politics* 20, no. 5 (2011): 668–686.

Eiser, J., J. van der Pligt, and R. Spears. *Nuclear Neighbourhoods: Community Responses to Reactor Siting*. Exeter: University of Exeter Press, 1995.

Endres, D. "From Wasteland to Waste Site: The Role of Discourse in Nuclear Power's Environmental Injustices." *Local Environment* 14, no. 10 (2009): 917–937.

Ethics Commission for Safe Energy Supply. *Germany's Energy Transition: A Collective Project for the Future*. Berlin, 2011.

Federal Ministry for the Environment, Nature Conservation and Nuclear Safety. "The Path to the Energy of the Future—Reliable, Affordable and Environmentally Sound." Accessed July 21, 2011. http://www.bmu.de/english/energy_efficiency/doc/47609.php.

Federal Ministry of Economics and Technology (BMWi). *Energy Concept*. Berlin, Germany, 2010.

Fischhoff, B. "Risk Perception and Communication Unplugged: 20 Years of Process." *Risk Analysis* 15 (1995): 137–145.

Friedman, S. "Three Mile Island, Chernobyl and Three Mile Island: An Analysis of Traditional and New Media Coverage of Nuclear Accidents and Radiation." *Bulletin of the Atomic Sciences* 67, no. 5 (2011): 55–65.

Fujino, S. "Fukushima Crisis: Nuclear Only Part of Japan's Problems." *BBC Online*, July 20, 2011, accessed August 5, 2011. http://www.bbc.co.uk/news/world-asia-pacific-14224409.

Gamson, A., and A. Modigliani. "Media Discourse and Public Opinion on Nuclear Power: A Constructionist Approach." *American Journal of Sociology* 95, no. 1 (1989): 1–37.

Gosden, E. "E.ON and RWE Scrap UK Nuclear Power Plans." *Telegraph*, March 29, 2012, accessed October 11, 2012. http://www.telegraph.co.uk/finance/newsbysector/energy/9173233/E.ON-and-RWE-scrap-UK-nuclear-power-plans.html.

Greenpeace UK. "Governments Nuclear Plans Declared Unlawful by High Court." Press release, February 15, 2007, accessed July 19, 2011. http://www.greenpeace.org.uk/media/press-releases/governments-nuclear-plans-declared-unlawful-by-high-court.

Hale, B. "Fukushima Daiichi, Normal Accidents, and Moral Responsibility: Ethical Questions about Nuclear Energy." *Ethics, Policy and Environment* 14, no. 3 (2011): 263–265.

Hendry, C. "The Road to Final Investment Decisions." Speech to the NIA Conference, July 6, 2011, accessed July 22, 2011. http://www.decc.gov.uk/en/content/cms/news/ch_speech_nia/ch_speech_nia.aspx.

Henwood, K., N. Pidgeon, K. Parkhill, and P. Simmons. "Researching Risk: Narrative, Biography, Subjectivity." *Forum Qualitative Social Research* 11, no. 1 (2010): article 20.

HM Chief Inspector of Nuclear Installations. "Japanese Earthquake and Tsunami: Implications for the UK Nuclear Industry Interim Report." London, UK: Office for Nuclear Regulation—An Agency of the Health and Safety Executive, 2011, accessed July 21, 2011. http://www.hse.gov.uk/nuclear/fukushima/interim-report.pdf.

Hoffman, S. "Negotiating Eternity: Energy Policy, Environmental Justice, and the Politics of Nuclear Waste." *Bulletin of Science, Technology & Society* 21(6) (2001): 456–472.

IAEA (International Atomic Energy Agency). "IAEA Update on Fukushima Nuclear Accident." *Fukushima Nuclear Accident Update Log*, April 12, 2011, 4:45, accessed April 12, 2011. http://www.iaea.org/newscenter/news/tsunamiupdate01.html.

IEA (International Energy Agency). "IEA Annual Report: World Energy Outlook 2010." IEA, 2010, accessed April 4, 2011. http://www.worldenergyoutlook.org/.

IEA (International Energy Agency). IEA, 2011, accessed July 22, 2011. http://www.iea.org/aboutus/faqs/nuclear/.

Ipsos MORI. "Strong Global Opposition to Nuclear Power." *Ipsos Global @dvisor Wave* 20, accessed July 25, 2011. http://www.ipsos-mori.com/researchpublications/researcharchive/2817/Strong-global-opposition-towards-nuclear-power.aspx.

Jasanoff, S. *Designs on Nature: Science and Democracy in Europe and the United States*. Oxfordshire: Princeton University Press, 2005.

Jewell, J. "Ready for Nuclear Energy: An Assessment of Capacities and Motivations for Launching New National Nuclear Power Programmes." *Energy Policy* 39 (2011): 1041–1055.

Kan, Naoto. "Press Conference by Prime Minister Naoto Kan." Press conference, Japan, July 13, 2011 (provisional translation), accessed July 20, 2011. http://www.kantei.go.jp/foreign/kan/statement/201107/13kaiken_e.html.

Kermisch, C. "Questioning the INES scale after the Fukushima Daiichi Accident." *Ethics, Policy & Environment* 14, no. 3 (2011): 279–283.

Klinsky, S., and H. Dowlatabadi. "Conceptualisations of Justice in Climate Policy: Synthesis Article." *Climate Policy* 9 (2009): 88–108.

Lynas, M. "Fukushima: Rationality versus Emotion in Policy-Making." Mark Lynas website, May 19, 2011a, accessed August 5, 2011. http://www.marklynas.org/2011/05/fukushima-rationality-vs-emotion-in-policy-making.

Lynas, M. "Fukushima's Lessons in Climate Change." *New Statesman*, March 17, 2011b, accessed October 10, 2012. http://www.newstatesman.com/asia/2011/03/nuclear-power-lynas-japan.

Masco, J. *The Nuclear Borderlands: The Manhattan Project in Post-Cold War New Mexico*. Princeton, NJ: Princeton University Press, 2006.

Meshkati, N. "Human Factors in Large-Scale Technological Systems' Accidents: Three Mile Island, Bhopal, Chernobyl." *Organization & Environment* 5 (January 1, 1991): 133–154.

McCurry, J. "Japan Cabinet Approves Fukushima Nuclear Compensation." *Guardian*, June 14, 2011, accessed July 25, 2011. http://www.guardian.co.uk/world/2011/jun/14/japan-approves-fukushima-nuclear-compensation.

Mogi, C. "Energy Policy Chaos Threatens Japan's Economy." *Reuters*, August 4, 2011a, accessed August 5, 2011. http://www.reuters.com/article/2011/08/04/us-japan-energy-idUSTRE7731GS20110804.

Mogi, C. "Japan's Nuclear Crisis Casts Doubt over Carbon Goals." *Reuters*, April 4, 2011b, accessed August 5, 2011. http://www.reuters.com/article/2011/04/04/us-japan-co-idUSTRE73233A20110404.

Moller, N., and P. Wikman-Svahn. "Black Elephants and Black Swans of Nuclear Safety." *Ethics, Policy and Environment* 14, no. 3 (2011): 273–278

Mombiot, M. "Why Fukushima Made Me Stop Worrying and Love Nuclear Power." *Guardian Online*, March 21, 2011, accessed March 10, 2012. http://www.guardian.co.uk/commentisfree/2011/mar/21/pro-nuclear-japan-fukushima.

Normimitsu, O., and K. Belson. "Culture of Complicity Tied to Stricken Nuclear Plant." *New York Times*, April 26, 2011, accessed August 4, 2011. http://www.nytimes.com/2011/04/27/world/asia/27collusion.html?pagewanted=all.

OECD Nuclear Energy Agency. "Public Attitudes to Nuclear Power." NEA No. 6859, OECD, 2010, accessed May 18, 2012. http://www.oecd-nea.org/ndd/reports/2010/nea6859-public-attitudes.pdf.

Office for Nuclear Regulation. "Japanese Earthquake and Tsunami: Implications for the UK Nuclear Industry, Interim Report." Merseyside: HM Chief Inspector of Nuclear Installations, May 18, 2011, accessed August 14, 2011. http://www.hse.gov.uk/nuclear/fukushima/interim-report.pdf.

Parkhill, N., Pidgeon, K. Henwood, P. Simmons, and D. Venables. "From the Familiar to the Extraordinary: Local Residents' Perceptions of Risk When Living with Nuclear Power in the UK." *Transactions of the Institute of British Geographers* 35 (2010): 39–58.

Parkhill, K., K. Henwood, N. Pidgeon, and P. Simmons. "Laughing It Off? Humour, Affect and Emotion Work in Communities Living with Nuclear Risk." *British Journal of Sociology* 62, no. 2 (2011): 324–346.

Petryna, A. *Life Exposed: Biological Citizens after Chernobyl*. Princeton, NJ: Princeton University Press, 2002.

Pidgeon, N. "Written Evidence on Public Attitudes to Nuclear Energy in House of Commons Science and Technology Committee 'Devil's Bargain? Energy Risks and the Public.'" *First Report of Session 2012–2013*, HC428. London: The Stationary Office, 2012: EV66–71.

Pidgeon, N., and B. Fischhoff. "The Role of Social and Decision Sciences in Communicating Uncertain Climate Risks." *Nature Climate Change* 1 (2011): 35–41.

Pidgeon, N., K. Henwood, K. Parkhill, D. Venables, and P. Simmons. "Living with Nuclear Power in Britain: A Mixed Methods Study." Cardiff, UK: School of Psychology Cardiff University, 2008a, accessed July 25, 2011. http://www.understanding-risk.org/docs/livingwithnuclearpower.pdf.

Pidgeon, N., R. Kasperson, and P. Slovic (eds). *The Social Amplification of Risk*. Cambridge: Cambridge University Press, 2003: 13–46.
Pidgeon, N., I. Lorenzoni, and W. Poortinga. "Climate Change or Nuclear Power—No Thanks! A Quantitative Study of Public Perceptions and Risk Framing in Britain." *Global Environmental Change* 18 (2008b): 69–85.
Porritt, J. "Why the UK Must Choose Renewables Over Nuclear." *Guardian Online*, July 26, 2011, accessed September 23, 2011. http://www.guardian.co.uk/environment/blog/2011/jul/26/george-monbiot-renewable-nuclear.
Rani, O. "It's Time Humans Heeded Nature Warning." *China Daily,* August 4, 2011, accessed August 4, 2011. http://www.chinadaily.com.cn/cndy/2011-03/19/content_12195463.htm.
Renn, O. *Risk Governance: Coping with Uncertainty in a Complex World*. London: Earthscan, 2008.
Rosa, E., and W. Freudenburg. "The Historical Development of Public Relations to Nuclear Power: Implications for Nuclear Waste Policy." In *Public Reaction to Nuclear Waste: Citizens' Views of Repository Siting,* edited by R. Dunlap, M. Kraft and E. Rosa, 32–63. Dunham, NC: Duke University Press, 1993.
Schneider, M., A. Froggatt, and S. Thomas. "The World Nuclear Industry Status Report 2010–2011, Nuclear Power in a Post-Fukushima World: 25 Years after the Chernobyl Accident." Washington, DC: World Watch Institute, April 2011: 7–8.
Shockley, K. "Fragility, Stability and Our Ideals Regarding the Well-Being of Others: Reflections on Fukushima Daiichi." *Ethics, Policy and Environment* 14, no. 3 (2011): 291–295
Shrader-Frechette, K. "Trading Jobs for Health: Ionizing Radiation, Occupational Ethics and the Welfare Argument." *Science and Engineering Ethics* 8, no. 2 (2002): 139–154.
Shrader-Frechette, K. "Fukushima, Flawed Epistemology and Black Swan Events." *Ethics, Policy and Environment* 14, no. 3 (2011a): 267–272
Shrader-Frechette, K. "Climate Change, Nuclear Economics and Conflicts of Interest." *Science and Engineering Ethics* 17 (2011b): 75–107
Slovic, P. "Perceptions of Risk." *Science* 236 (1997): 280–285.
Spence, A., D. Venables, N. Pidgeon, W. Poortinga, and C. Demski. "Public Perceptions of Climate Change and Energy Futures in Britain: Summary Findings of a Survey Conducted in January–March 2010 (Technical Report—Understanding Risk Working Paper 10–01)." Cardiff: School of Psychology, Cardiff University, 2010, accessed July 25, 2011. http://www.understanding-risk.org/docs/final_report.pdf.
Stanley, A. "Just Space or Spatial Justice? Difference, Discourse, and Environmental Justice." *Local Environment: The International Journal of Justice and Sustainability* 14, no. 10 (2009): 999–1014.
Stirling, A. "Limits to the Value of External Costs." *Energy Policy* 25, no. 5 (1997): 517–540.
US Nuclear Regulatory Commission. "Inspections at U.S. Nuclear Plants Prompt Corrective Actions." Press release, Washington, DC, May 13, 2011.
van de Poel, I. "Nuclear Energy as a Social Experiment." *Ethics, Policy and Environment* 14, no. 3 (2011): 285–290.
Welsh, I. *Mobilising Modernity: The Nuclear Moment*. London: Routledge, 2000.
The White House. "Blueprint for a Secure Energy Future." Washington, DC, March 30, 2011.
Wynne, B. "Misunderstood Misunderstandings: Social Identities and Public Uptake of Science." *Public Understandings of Science* 1 (1992): 281–304.
Wynne, B. "May the Sheep Safely Graze: A Reflexive View of the Lay-Expert Knowledge Divide." In *Risk, Environment and Modernity*, edited by S. Lash, B. Szersynski and B. Wynne, 44–83. London: Sage, 1996.
Zonabend, F. *The Nuclear Peninsula*. Cambridge: Cambridge University Press, 1993.

9 The Search for Energy Security after Fukushima Daiichi

Jim Falk

This book includes amongst its objectives the important question: Should we continue with nuclear power as a key technology for a low carbon future? (see Chapter 1). Governments have a key role to play in answering this question, and often the way they address it is in terms of the potential of nuclear power to maintain or increase energy security. The question of what is the best way to assure low carbon energy security is a pressing one. However, when addressed by policy makers the analysis is often set in terms where technical considerations (including, for example, capital and running cost, engineering risk analyses, construction time, grid stability, infrastructure requirements, and a range of similar factors) are given priority.

The Fukushima crisis gives us reason to question this formulation, precisely in the way in which the field of science, technology and society (STS) so often invites us to. That is, it suggests we go beyond traditional technical perspectives and see how they are connected to the social. A key tool in this is the concept of the sociotechnical system, utilized in a variety of STS analytical approaches. Sociotechnical systems, as Miller et al. (2008: 3) put it, are 'coupled' so as to 'link human and social values, behaviour, relationships, and institutions to science and technology . . . [that] are central to understanding the nature and dynamics of sustainability problems and solutions'. This chapter will consider the implications of the Fukushima Daiichi reactor accident for the future role of nuclear power in energy security from a sociotechnical system perspective.

In examining the preceding topic it is useful to consider nuclear power as one example of what is a more general class of large scale technologically sophisticated systems that rely on the latest engineering technology to contain a potent hazard. Examining some of the common features of such systems suggests that it is useful to go beyond the more instrumental considerations within which considerations of energy security are frequently couched. In particular, it is important to take into account the social dimensions of the system within which the technical aspects are embedded. Doing this, it is argued, is central to understanding the implications of the Fukushima crisis for the potential role of nuclear power, at a time when planning for a low carbon future is needed. To introduce this argument it is useful to briefly recapitulate a story which may well seem hauntingly familiar.

A TALE OF TWO SYSTEMS

On an otherwise calm evening the operation of one of the most massive, complicated, and sophisticated technological systems in the world was suddenly and devastatingly disrupted. Powered by the latest energy conversion system it was operating at close to full power. To those who relied on it everything seemed to be working predictably and safely, but this sophisticated system was soon confronted by a well-known if infrequent natural event. Although protected by elaborate defenses designed to make a system breach vanishingly improbable, over a brief time a sequence of events was set in motion that critically undermined all the designed protection. Whilst other factors contributed, it was ultimately an inward surge of seawater that prevailed. Barriers supposed to hold back the incoming flood were progressively overcome until a tipping point was passed. The incoming water intruded into, shorted out, and rendered inoperable the electric systems. Then, to the horror of onlookers, venting plumes of steam driven by its hot inner core, the massive and seemingly unbreakable outer shell of the system fractured to the sound of violent explosions. As it succumbed to the forces of the sea, the system was irretrievably wrecked, with the lives of thousands of people catastrophically affected.

The preceding is actually an account of what transpired on April 14, 1912, when the ship *Titanic* on its maiden voyage struck an iceberg (sinking two hours and 50 minutes later). Perhaps peculiarly, it also applies to the more recent events at the Fukushima Daiichi reactor complex initiated by an undersea high intensity earthquake and subsequent tsunami. This occurred on March 11, 2011, at 2:46 p.m., 99 years after the sinking of the *Titanic.*

There are, of course, many differences between the circumstances of the *Titanic* and Fukushima nuclear power plant. The two technologies had entirely different purposes and sources of energy (burning coal versus nuclear fission in each case to make steam). Whilst the *Titanic* was brand new, the Fukushima Daiichi reactors were near the end of their operating life when the tsunami hit.

The *Titanic*'s designed protection failed catastrophically in three hours as it sank after striking the iceberg, whilst at Fukushima the earthquake and tsunami led to a more slowly played out breach of the system protection leading to successive releases of radiation (Kang 2011). According to one estimate, by April 27 approximately 55% of the fuel in reactor unit 1 had melted, along with 35% of the fuel in unit 2, and 30% of the fuel in unit 3. On April 12, the Japanese government finally conceded that this was a level 7 accident on par with what occurred at Chernobyl (Grier 2011). The human impacts in each case were also cruelly if differently damaging. But many haunting similarities remain:

1. **Each was a large, technically sophisticated and technologically complex system.** In 1912, the *Titanic* was the largest ship in the world. In 2011, Fukushima (including the six boiling water reactors at the Daiichi site, *together* with the four reactors at the neighboring Daini site

1.5 kilometers to the south) was the largest nuclear power complex in the world (Mohrbach 2011).

2. **Each sought to contain a potentially catastrophic threat.** The *Titanic* was designed to keep the danger of the sea out, protecting the passengers within. The comparable task at Fukushima Daiichi was to keep the nuclear hazards in, protecting the population outside.
3. **Each was protected by designed 'defense in depth'.** The *Titanic* and the Fukushima Daiichi reactors were protected by multiple defenses against accidents—known as 'defense in depth'—intended to make a catastrophic accident vanishingly unlikely. This included, in each case, two layers of containment.

The *Titanic* had a double-bottomed hull constructed of one-inch-thick steel plates, intended to provide twice the normal strength in a collision, and also an inner wall against seawater if the outer hull was breached (McCluskie et al. 1998:107–109). Further, the lower part of the ship was divided into 16 compartments, capable of being sealed by electrically operated doors to keep the ship afloat even in a collision where two compartments were completely flooded. Complementing these were 57 additional watertight compartments, of which 43 were contained within the double bottom hull (Brown 2001:119), which the owners described as 'practically making the ship unsinkable' (McCluskie et al. 1998: 110).

Fukushima Daiichi also had a double walled containment, the outer shell made of reinforced concrete, the inner of steel. An earthquake would automatically trigger insertion of control rods halting the nuclear reaction. The ground level of the reactor units was 10 meters above sea level providing protection in the unlikely event of a tsunami wave (Mohrbach 2011: 4). This conformed to the Japanese Nuclear Safety Commission guidelines that utilities take tsunami danger into account, a modest requirement because the Commission also had concluded: 'Even for a nuclear plant situated very close to sea level, the robust sealed containment structure around the reactor itself would prevent any damage to the nuclear part from a tsunami, though other parts of the plant might be damaged. No radiological hazard would be likely' (Perrow 2011).

4. **For each, key management approaches were dismissive of the dangers faced.** The Titanic sailed through a known ice field at close to full speed with inadequate lookouts for the conditions and without changing course (which it could have) to avoid the danger (Brown 2001: 5). The speed may have in part reflected pressure on the captain from the managing director of the shipping line, J. Bruce Ismay, to outshine competitors by arriving earlier than scheduled in New York (Brown 2001:192–193).

At Fukushima Daiichi, a decision to delay venting hydrogen (accumulating in the reactors from overheated fuel) contributed to the explosive destruction of the three reactor outer containment structures (Powell and

Takayama 2011: 5). Another had been the much earlier decision to build one of the world's largest nuclear complexes on a coastline where huge earthquakes had been a regular fact of life.

5. **Each had prior warning of the risk of the natural event that ultimately overwhelmed it.** In the case of *Titanic*, officers received a series of warnings from other ships of the ice field ahead. The ship *California* (around 32 kilometers away) had been so shaken by an ice collision that the captain had stopped the ship for the night. Its warning did not get to the *Titanic*'s officers, largely because the importance of this was not made clear (Brown 2001: 191–192).

At Fukushima Daiichi, prior warning was a much slower affair, like the incident itself. Japan had experienced smaller earthquakes that had generated tsunamis around the design height. For example, the west coast of Japan had experienced a nine-meter-high tsunami in 1993 (Perrow 2011). Multiple statements of concern about the modest protection against tsunamis had been made by a member of the national parliament, and a member of the nuclear safety board had resigned in 2006 over lack of attention paid to such risks in nuclear design (Perrow 2011). As recently as 2008, it is reported that the company TEPCO (Tokyo Electric Power) had revaluated tsunami risk, with simulations showing waves as high as 15 meters (almost exactly the height reached in 2011) but the company leadership ultimately chose to disregard this as unreliable (Powell and Takayama 2011: 2–3).

In the cases of both Fukushima Daiichi and the *Titanic* there appears to have been a high level of complacency in relation to the dangers faced, resulting in inadequate preparation to deal with them.

6. **In each, many vital elements of the designed safety measures worked as planned but nevertheless did not prevent catastrophic failure.** Just prior to contacting the iceberg, *Titanic*'s electromagnetically triggered watertight doors began to descend, as ordered, sealing off the compartments shortly thereafter (Brown 2001, 77–78). *Titanic*'s engines were also stopped and engineers took other protective actions.

At Fukushima Daiichi the electromagnetically triggered control rods dropped as planned and closed off the nuclear chain reaction at the moment of the earthquake.

7. **In each, the circumstances of the accident and consequent situation of the wreckage left open to debate the exact chains of causation.** *Titanic* lies under crushing water pressures 4,000 meters below the surface. At Fukushima Daiichi, high radiation in melted cores (which at least in one reactor melted through the base), destroyed the

outer containment and damaged spent fuel ponds, profoundly hindering thorough inspection.

8. **In each, reliance on a massively powerful heat source posed a particular threat as containment failed.** In the *Titanic,* engineers and stokers worked frenetically to prevent a steam explosion wrecking the ship by letting off steam, removing fuel from the furnaces, and raking out the coals. However, this did nothing to quench the incoming flood, and merely deferred the inevitable outcome including explosions observed in its final moments (McCluskie et al. 1998: 312).

At Fukushima Daiichi, control rods stopped the immediate chain reaction but could do nothing about the heat still being released in great amount by radioactive decay within the fuel rods. Continuous cooling was required but failed ultimately, causing explosions and radiation release.

9. **Each demonstrated physical inadequacy in the face of the known dangers.** In the case of the *Titanic,* metallurgical study of material salvaged from the shipwreck suggests that substandard rivets were used that would not have withstood well the stresses created as the ship flexed on contact with the iceberg (Garzke et al. 1994, Garzke et al. 2000, K. Smith et al. 2005). The sealed bulkheads proved inadequate as water poured in through the damaged hull, because as the water rose it could flow over the deck above the compartments. As the ship bow sunk, the compartments progressively filled from above, assuring the *Titanic*'s doom. Had the bulkheads extended further above the water line, as was done with later designs, and been fully watertight once the doors were closed, catastrophe might well have been averted.

In the case of Fukushima Daiichi, the reactor was not high enough above sea level, and there was inadequate protection from flooding provided for crucial power systems. Power cables and backup generators in the lower part of the plant could have been placed in a situation that was not vulnerable to a wave of this magnitude, which was a known possibility.

10. **In each, the poor design and management decisions compounded the problems.** For the *Titanic* there is evidence that the damage to its inner bottom shell was such that pumping, which commenced after the collision, may have been able to keep the ship afloat until help arrived (about two hours after the *Titanic* sank). However, that potential would have been undermined by a fateful decision, possibly under pressure from the managing director of the company (who was clearly concerned about the company's

reputation). Testimony at the inquiries suggests that after striking the iceberg, but before fully assessing damage, the ship restarted its engines and steamed on at low speed for about six minutes (Brown 2001:128–146). It is likely this would have dramatically increased the inrush of water through the damage to the inner hull.

At Fukushima Daiichi, the designers had known that in the event of a loss of cooling, the reaction between the steam and the zirconium cladding of the fuel rods would release hydrogen. Nevertheless, the decision to vent the hydrogen was delayed and the outer containment structure was not strong enough to withstand the resultant explosions. The situation of spent fuel in tanks above the reactors made easier the routine processes of removing fuel rods from the reactors but left a very large amount of highly radioactive fuel extremely vulnerable to loss of cooling, leading to fires, further radiation release, and danger of collapse of the supporting structure.

11. **In each there was lack of preparation for a catastrophic failure in the safety systems.** The *Titanic* notoriously carried only enough lifeboats for a little over one third of its passengers. Whilst exceeding the then Board of Trade requirements, this reflected both outdated regulations and a view that modern ships would sink slowly enough for other ships to come to their rescue (McCluskie et al. 1998: 136).

At Fukushima Daiichi, despite decades of operation and several extremely serious prior nuclear reactor accidents around the world (for example, at Chernobyl and Three Mile Island), preparation for the events as they unfolded was almost nonexistent. According to one report, the company (TEPCO) that ran the reactors at one stage proposed to abandon the site altogether but was dissuaded by then Prime Minister Kan (Powell and Takayama 2011: 2). In the event, the company relied on a ramshackle set of interventions—for example, fire engines, helicopters, and even riot control vehicles—to try to get sufficient water into the damaged reactors and fuel pond to reduce the level of meltdown already taking place.

12. **In each, the final safety option, to remove endangered people from the reach of the threat, was undermined by extraneous considerations.** Evacuation was delayed and inhibited by a lack of transparent communication as authorities sought to find a balance between news that would cause panic and reputational damage, and prompting evacuation that might save lives.

Many of the *Titanic*'s crew were not fully informed and initially told passengers to return to their cabins. Many passengers were advised to don life jackets but were not advised to proceed to the deck to board lifeboats. The lack of impact on the ship's motion when it struck (a faint shudder as the ship ground against ice), and the restarting of engines and continuation of life as usual, provided a reassuring atmosphere in which, until too late, even many first class passengers believed they would be safer in the ship than in lifeboats.

At Fukushima Daiichi an apparent desire by the plant's owner, TEPCO, and the government, to keep damaging information to itself, rather than cause panic and feed more general concern over the nuclear industry (cf. Morita et al. in this book), created delays. It was not until April 20, more than five weeks into the accident, that the area within 20 kilometers of the site was declared a 'no go zone,' shattering hopes that there could be a return to normal soon (Japan Times 2011). Vulnerable populations often had insufficient information about their radiation exposure. Information released by TEPCO, the government, and officials from the regulatory agencies was frequently conflicting and insufficiently detailed. In order to rectify this, a concerted effort was made by many civic interests to produce alternative radiation maps (see Morita et al. and Kera et al. in this book), as well as Greenpeace, which deployed its own team to monitor radiation (Hongo 2011). Significant hot spots of radiation were found in surrounding communities, ultimately forcing further evacuation.

13. **In each, impacts spread unevenly across the affected population.** Impacts of such accidents percolate across layers of the sociotechnical system in which they occur. Workers onsite are affected differently than people living nearby. Over time, many more may find themselves affected but with different vulnerabilities for different subgroups.

In the *Titanic,* access to the deck for different classes of passengers reduced from first to third class, with correspondingly lethal consequences (Hall 1986: 687). Because of insufficient number of lifeboats the crew had to resort to 'women and children first' in order to ration seats in them. Some members of the crew (for example, the engineers below deck) also had much reduced chances of escape. Beyond that, the potentially affected groups included the relatives of those who died, traumatized survivors, and the shipping industry itself, although in the wake of this event it retained its momentum.

Over 1,500 people perished on the *Titanic.* At Fukushima Daiichi, apart from some workers being exposed to high levels of radiation (see also Pritchard in this book: 124 and 128), estimates of the number of lives shortened by the accident will depend on technical arcana of radiation population dose models; disputes over the impacts of low levels of radiation (Beyea 2012); estimates of the amount, rates, and types of radiation omitted; considerations of impacts of local 'hot spots'; and much else. One report estimates that the Fukushima Daiichi accident will result in a maximum of 1,300 deaths and 2,500 cases of illness from radiation alone (excluding increased risk for the 20,000 workers at the plant). Added to these are nearly 600 deaths in the region already certified as 'disaster-related,' many of which were associated with the evacuation (Ten Hoeve and Jacobson 2012).

The varying vulnerability of subpopulations also reflects their economic capacity to evacuate, and having done so, to thrive. Vulnerability to radiation varies across subpopulations according to gender, age, and health. On April 25, the Japanese government announced an increase to permitted doses of radiation (from the International Commission on Radiological Protection maximum dose for the public of 1 mSv/y to 20 mSv/y), rather than order more extended evacuation (Buongiorno et al. 2011). This was particularly controversial in relation to children who are more vulnerable to radiation than adults. In some cases, levels of radiation permitted to Japanese residents were above the levels at which the Soviet authorities had ordered protective evacuation around Chernobyl (McKenzie 2011).

Beyond health impacts was the traumatic disruption to the lives of the affected. The legacy of *Titanic* included psychological impacts on survivors and those who mourned lost loved ones. In the Fukushima Daiichi aftermath there is an unfolding legacy of disrupted businesses, abandoned communities, and profoundly disrupted lives (e.g. Osang 2011). For both *Titanic* and Fukushima Daiichi, access to safe options differed significantly according to affluence.

14. **For each, there was a continuing struggle by various key interests to cast the history of what had happened in a favorable light.** At the inquiries in New York and London into the sinking of the *Titanic* (Ottmers n.d.), the White Star Shipping Company officers who testified were keen to present the company in as good a light as possible. This included dodging the recollections by other survivors that the ship broke in half at the surface (subsequently found to be true), justifying moving at near full speed through the ice field as normal practice (Brown 2001), and side-stepping the fact that the ship had been put underway again after the collision with the iceberg. As the highest-ranking surviving officer, Second Officer Lightholler mentioned in his memoirs but not at the hearings, 'in London it was necessary to keep a firm hand on the whitewash brush' (Hall 1986: 689, Brown 2001: 3).

With the benefit of 100 years of hindsight it is clear that many issues were brushed over. Key attention was directed to the lack of adequate lifeboats and other safety features that could be easily enough rectified.

Critically we do not have the benefit of hindsight in relation to the Fukushima Daiichi accident. There is a well-known closeness between the ruling party, Japanese government agencies, and the nuclear industry (Moe 2012), and despite the government's announcement on December 16, 2011, that the plant was now in 'cold shutdown', as reported five months later,

> as workers today at the nuclear power plant, will tell you, is this: it wasn't true then, and it's still not true today. "The coolant water is keeping the reactor temperatures at a certain level, but that's not even near the goal [of a cold shut down]", says an engineer working inside the plant. "The fact is, we still don't know what's going on inside the reactors". (Powell and Takayama 2011)

More broadly there has been evidence of efforts at government level to play down the severity of the ongoing crisis. This includes reports of measures by Japanese agencies to control information 'harmful to public order and morality' (Segawa 2011), and agreement between the US and Japanese governments to play down the severity of events. A government worst-case scenario requiring the evacuation of Tokyo was kept secret (Japan Times 2012). Finally, the operation of the 'whitewash brush' could be seen once again in London with decisions by the UK government to launch a public relations campaign to play down any adverse implications for new reactors in Britain (Edwards 2011).

SIMILARITIES AND SYSTEM SCALE

Summarizing the commonalities exhibited in this 'tale of two systems', we might say that each was representative for its time, of pioneering technology, grandiose ambition, and herculean construction effort. Each was supposed to be near invincible. And each was defeated by the simple force of water. As Charles Dickens (1859: 1) wrote 53 years before *Titanic* met its fate, for the corresponding societies: 'It was the best of times, it was the worst of times. It was the age of wisdom. It was the age of foolishness . . .', to which for Fukushima could be added the near contemporaneous remark by Jean-Baptiste Karr (1849: vi) that 'plus ça change, plus c'est la même chose' ('the more it changes, the more it is the same thing').

Despite the vast difference in technology, purpose, and time, an eerie commonality can be found between the events around the *Titanic* and Fukushima Daiichi. Indeed, similar commonalities could also be easily found between them and other major nuclear accidents (for example, Chernobyl and Three Mile Island). In each, accidents considered so remote as to be vanishingly unlikely took place through a complex skein of human and technical missteps in which, even if each may have been previously considered individually, in combination they were neither anticipated nor adequately prepared for. Similarity in responses (from reluctance to evacuate to lack of transparency about the unfolding events) is also readily identifiable (Falk 1982: 29–42, Meshkati 1991).

This raises the simple question: Why, when separated so far in time, space, and purpose do these disparate examples show so many aspects in common? The answer may well be complicated, but almost certainly partly reflects deeper commonalities. One commonality is their '*high scale*' and design to protect against '*high consequence vulnerability*'. Here, 'high scale' is intended to convey a combination of large scale and high technology. The design and construction of these systems was not only physically big, but utilized the most sophisticated engineering available. Each thus presents an example of what are sometimes called Large Scale Technical Systems (e.g. La Porte 1989). 'High consequence vulnerability' is intended to convey that these systems face a potential for catastrophic failure. Sophisticated design and technology is utilized in an effort to render such a failure vanishingly unlikely.

But, the systems considered here were not restricted to inanimate objects. They were 'sociotechnical systems' composed not only of objects but also people, institutions, beliefs, interests, and contested claims (e.g. Bijker et al. 1987). The commonalities between the different accident's sequences considered could be found spreading widely across sociotechnical systems incorporating not only the central ship or reactor but also the workers and management, the inhabitants and their relatives, the industry and its interests, and the connections in government and civil society. The commonalities turned out to be particularly strong in these more human aspects. In each case, the road to failure, even if commencing from one catalyzing causal incident, was paved with many events. Generally, the crisis was created by a 'perfect storm' of unexpected physical events, human and design errors, technological failures and consequent physical, economic and social repercussions (e.g. Meshkati 1991).

It is worth considering some general aspects of these features of scale. First, investment in and construction of 'high scale' systems (such as the *Titanic* and the Fukushima Daiichi reactors) is possible only for very large companies. Being very large also vested in these companies a competitive advantage over smaller ones. Over several centuries there has been a trend toward ever larger scale economic organization. This includes the emergence of larger and larger corporations to the point where today the economic capacity of some exceeds that of many nation states. Even 100 years ago, when the *Titanic* was manufactured, economic organizations of the scale of the White Star Line had a close relationship to national governments and their agencies. Further, the ambitions of such corporations could reinforce and be reinforced by those of governments.

The *Titanic* represented a further step in the pattern of development of ever larger ocean liners in the competition between shipping lines. These large liners were also symbolic in a competition between England and Germany for civilian and military marine supremacy (Howells 1999: 19). Similarly the reactors constructed at Fukushima Daiichi were part of the continuing effort to establish greater energy security for Japan, particularly after the Arab oil embargoes of 1967 and 1973–1974 (Sawa 2012).

More broadly, the nuclear industry has gained status from its historical co-development with nuclear weapons. Nuclear weapons have, in turn, played important roles in the regional and global power achieved by the nuclear weapons states. Nuclear power has acted as an adjunct to, proxy for, or at times, the route towards nuclear weapons (e.g. in the former USSR, USA, UK, France, China, Iran, India, Pakistan, and North Korea).

The 'high scale' of nuclear energy production has also positioned it well as a symbol of national capability. Governments have often had to collaborate closely with large corporations to achieve that objective. For these reasons, from the beginnings of the nuclear industry, there has been a close interrelationship between the development of the commercial technology, actual or potential military applications, and the state (e.g. Camilleri 1984). That close relationship had made it more likely that governments would be protective of the industry, thus shaping in part their responses when potentially catastrophic nuclear accidents are in play.

On the morning of the news of the loss of the *Titanic,* three cardinals wrote to President Taft, expressing: 'profound grief at the awful loss of human lives attendant upon the sinking of the steamship Titanic', and 'the hope that the lawmakers of the country will see in this sad accident the obvious necessity of legal provisions for greater security of ocean travel' (O'Connor 2012).

Similarly, each of the major nuclear disasters has created a widespread response both in the demand and provision for greater safety (e.g. Juraku, and Schmid, in this book) and indeed, for a more critical response to any further development of the use of nuclear power. The sociotechnical system in which the events at Fukushima are a key actor then reaches out to embrace not only the nuclear power industry, but the role of energy throughout the entire economy. And the contest over its implications, in a similar way to the sinking of the *Titanic,* reaches out to raise the whole question of nuclear power's role in relation to energy security.

SOME IMPLICATIONS FOR 'ENERGY SECURITY'

'Energy security' can be thought of in several ways. One useful approach, as suggested previously, is to consider it in the context of sociotechnical systems. But this is not the orthodox way to talk about energy security. It is most usually invoked as a component of national security (Falk 2012). The Australian Government's Department of Energy, Resources and Tourism (2009: 5) has defined energy security (for Australia) as 'the adequacy, reliability and affordability of energy supplies'. In an effort to take on board the relationship between energy generation and climate change (Scott 1994: 35), the International Energy Agency (2012a) recasts this concept in terms of energy's 'uninterrupted physical availability at a price which is affordable, while respecting environment concerns'.

The twin criteria of adequacy and reliability have often informed a claim that the use of nuclear power is essential for ensuring energy security. This breaks down into a claim that nuclear power is necessary:

- to provide enough energy to support national economic and social activity (adequacy);
- with minimal interruptions to supply (reliability);
- commensurate with maintaining economic activity and competitiveness and continued investment in the energy sector (affordability).

Consistent with this, as the International Energy Agency (2011: 1) notes, 'events such as those at the Fukushima Daiichi nuclear power plant and the turmoil in parts of the Middle East and North Africa (MENA) have cast doubts on the reliability of energy supply'. Yet, warns the Agency (which consistently tends to support nuclear power expansion), if the world embarks on a path of doing without expanding nuclear power this will 'put additional upward pressure on energy prices, raise additional concerns about energy security and make it harder and more expensive to combat climate change'.

Energy security here is expressed not only in national but also in instrumental terms. That is, security is taken to be a technically determinable condition, objectively defined in terms of firm criteria (even if establishing whether the criteria are met in any given circumstances could be a matter of considerable debate). Here, any technology that meets these criteria can be considered to satisfy the objectives of security. This approach to energy security, however, is confronted by several factors, including that:

1. **need** for security in energy is not solely manifested at the scale of nation states. Access to energy is a requirement for all communities however large or small.
2. **supply** of energy is not just a national matter. Coal furnaces, oil, nuclear reactors, and solar cells are all produced in a global production system and often consumed in a completely different place from which they are produced.
3. **mitigation** of greenhouse gas emissions associated with energy production is notoriously a global problem, even though the impacts and contributions to climate change are distributed unequally (and inequitably) across the planet. Conversely, **adaptation** is usually grounded in local action.
4. **impacts** (whether physical or social) of energy generation are distributed unequally within states. Whether for coal or nuclear power (as illustrated at Fukushima) local communities can suffer a disproportionate burden of risk and impact in relation to benefit.
5. **access** to energy is not distributed equally even in small communities. The benefits are usually distributed according to capacity to pay for electricity.

Thus energy security is likely to mean very different things to different communities at various scales. This raises the possibility that the lessons drawn from Fukushima may well be very different depending on the social and physical location and scale at which they are drawn. These conclusions will also, in part, be shaped by what we will call here 'boundaries to choice'.

BOUNDARIES TO CHOICE

Here 'boundaries to choice' will be taken to mean boundaries to what are understood to be feasible options for action in relation to perceived needs. These boundaries may be contested and change over time but at any moment they constrain what is understood to be feasible and thus what changes are initiated.

At the time of the sinking of the *Titanic*, there was no available feasible alternative to the use of shipping in intercontinental transport. After the accident, therefore, the focus was on ensuring that future and current ships were safer. Changes included sufficient lifeboats to seat everyone on board, higher bulkheads that created completely sealed compartments, double sides extending above the tank top deck, constant monitoring of ship's radio, greater caution in ice fields, and much else (Brown 2001: Appendix 2).

The circumstances of the nuclear power industry in the wake of the Fukushima crisis were different. The understanding of energy alternatives, whilst contested, was rapidly evolving. As a special Intergovernmental Panel on Climate Change (2012) report showed, a raft of energy alternatives was on offer or under development. Feasible future energy proposals might contain nuclear components or omit them (e.g. Elliston et al. 2012: 1, Moriarty and Honnery 2012). Nowhere were these alternatives more alive than in Germany, where a decision was announced to phase out all use of nuclear power by 2022, coupled with an ambition to become a world leader in renewable energy and energy conservation industries. Switzerland and Belgium also announced phaseouts (for 2025 and 2034, respectively) whilst Japan announced its intent to reduce dependence on nuclear power (International Energy Agency 2012b: 24).

The growing recognition that there were available alternatives resonated with the long-standing controversy over the safety, desirability, and other implications of use of nuclear power—a powerful controversy which had evolved with major ramifications in many countries, since its beginnings in the 1970s (Falk 1982).

ENERGY SECURITY, TECHNICAL CONTROVERSY, TRUST, AND SCALE

In the six years from 1972 to 1978 (the year before the Three Mile Island accident), the world nuclear industry's expectations of the amount of nuclear power to be generated in the year 2000 dropped by 72% from 3450 GWe (gigawatt-electric) to 728 GWe (Falk 1982: 23–27). The effect of Three Mile

Island was to further constrain an industry whose expectations were already in substantial decline. As at March 2012, actual world nuclear generating capacity was only 370 GWe (NEI 2012) and many reactors were near the end of their operating lifetimes.

As I argue elsewhere (Falk 1982, 2011), a key reason for the dramatic decline in nuclear expectations in the 1970s was citizen concern. This concern translated into pressure for a steady stream of changes to the design of reactors in an effort to improve safety (Bupp et al. 1975). Especially in the USA, citizen concern led to increases in capital cost associated also with tightened regulation, increased time to license, construct, build, and commence operation of new nuclear reactors and increased capital carrying costs. Each successive reactor incident led to increased opposition to nuclear power especially where new reactors were being proposed. As a result, citizen concern played a significant role in internalizing costs associated with the risks of nuclear power rendering it economically uncompetitive in many locations. Over the following decades, whilst the various factors mentioned above would affect costs differently in different places, later studies continued to show that the costs of nuclear power were highly sensitive (not infrequently fatally for the industry) to the safety requirements of the regulatory environment (e.g. Koomey and Hultman 2007).

Throughout the long but contested process of deployment of nuclear power across many countries, there has been pressure on the industry to innovate in order to deflect particular safety concerns. Whilst this may be a lesser pressure where it can be insulated from democratic process, in many countries that pressure has forced a process of continuous innovation. But, as already noted, this has led to an increase in capital costs. Even in France, where a strong alliance between government, bureaucratic agencies and the nuclear industry was able to hold innovation at bay as a matter of policy, Komanoff (2010) has shown that as soon as the program moved beyond that the cost rose to twice as much per GWe as for the previous 54 reactors already built. Thus while many renewable energy technologies (for example, solar, wind, biofuel) can be constructed as large numbers of small units and thus enjoy economies of scale from mass production, nuclear power has shown no such tendency.

Prior to the events at Fukushima Daiichi a similar situation to that existed for the nuclear power industry internationally prior to the Three Mile Island accident (Falk 2011). As Schneider et al. (2011: 7) reported, in most countries the capital costs of nuclear reactors were rapidly increasing, greatly undermining their economic competitiveness (unless supported by increased subsidies).

This situation where the economics of nuclear power were delicately balanced at best, and where increased concern might add to costs, was reflected clearly in a postscript to a report by the Massachusetts Institute of Technology Study Group (2011 Postscript: xv), hastily added in the wake of the Fukushima crisis. It noted that as a result, 'costs are likely to go up' because of new safety design requirements; 'the relicensing of forty year old nuclear plants for another twenty years of operation will face additional scrutiny' and some licensing extensions already granted may 'be revisited'; and the entire spent fuel management system 'is likely to be revaluated'. In concluding, it noted: 'How these and other post-Fukushima issues are resolved will have major implications for the future of nuclear power '.

The role of nuclear power in energy security is thus confronted by a series of social factors that will likely greatly affect its future. This reinforces the importance of examining the objective of energy security within the broader sociotechnical energy system. In particular, two things already mentioned seem particularly relevant:

1. What is understood as energy security will be affected by perceived boundaries to choice amongst energy options. These may seem different in different communities, but, at the time of Fukushima in many communities these perceived boundaries to choice seemed to be expanding.
2. What is understood to be economic and practical may be altered in the light of community understandings. In relation to the possible contribution of nuclear energy, community beliefs may reshape the costs of construction. Similarly, the impacts of regulation by government on the industry can be changed across a vast range of factors; from allowed radiation levels, requirements for managing nuclear waste, measures to avoid nuclear weapons proliferation, or safety improvements to reduce the likelihood of accident sequences of the sort revealed at Fukushima.

In short, what can serve as an economically viable and politically acceptable contributor to energy security is not merely a technical but also a social and economic question, whose answer may be profoundly affected by community sentiment. In particular, I argue, that sentiment will be shaped by *trust*. Dictionary definitions include in the connotations of trust, things, people, or institutions in which confidence is placed thus providing a basis for 'reliance, faith or hope' and dependability for 'competent, expert, honest, or safe behavior' (McKnight and Chervany 1996: 20–21, Webster On-line Dictionary 2012).

Trust is thus a different concept from *risk*, the term often used in engineers' assessments in proposals to build nuclear reactors. Unlike trust, risk is often allocated a numerical weighting through the calculation of probabilities, or estimates of likelihood. Trust goes beyond risk, for example, in raising the question of whether this or that risk assessment is to be trusted. In this sense, energy security is not simply an objective condition. It is a product also of multiple views, concerns, and judgments (however reached) made across communities. Institutions that fail to meet the collective requirements of those community judgments will fail to provide the required sense of 'security' and thus will find it harder to garner the trust of that community. One of the most insidious problems for any part of the energy industry is to build the required levels of community trust. This is made particularly difficult by its high consequence vulnerability and high scale.

High consequence vulnerability requires particularly high standards to be met for trust to be won. Major accidents highlight the vulnerability, draw attention to the potential for safety flaws, and make all the harder the task of building confidence in the effectiveness of government and industry responses. Trust can be undermined not only by the failure to meet technical requirements, but also by perceived deficiencies in the speed, transparency, and honesty with which industry and government respond to safety incidents.

The high scale of the nuclear industry also makes the task of building trust harder. Its scale involves widely dispersed activities spread from the regions where the reactors themselves are situated to the cities where the bulk of the electricity produced is consumed. One challenge for the industry is that benefits and risks can be seen to be inequitably distributed with risks suffered in the regions where reactors or waste disposal is sited, to provide power enjoyed primarily in the cities. In the past this has led to intensification of opposition in the regions (Falk 1982: 172–199).

Further, because of the high scale, the nuclear industry's management and regulators are usually located in major cities, close to government centers, and far from many of the communities that are party to its activities. A barrier of technical complexity compounds this potential for alienation of the industry's management from user communities (see also Hindmarsh in this book: chapter 4). This can make much more difficult the task of gaining trust, especially in times of crisis. Reduced trust may appear as decreased confidence that industry promises can be relied on, needed information will be honestly and quickly presented, and design for safety will trump considerations of profit.

Following the Fukushima crisis, the physical and social impacts (such as radiation exposure on the one hand and disrupted lives and lack of transparency on the other) seem to have undermined trust, especially in Japan (see also Hara in this book). Naoto Kan, prime minister at the time, said: 'We should seek a society that does not rely on nuclear energy' (Yamaguchi 2011). All reactors have been closed; at least for the time being (New Scientist 2012), and there is a very real prospect that for at least for some of them, local opposition may render that state permanent. More generally, the role of nuclear power in the perceptions of what constitutes energy security in many countries appears to have changed quite profoundly in the wake of the Fukushima crisis.

Thus, for example, a post-Fukushima GlobeScan (June–September 2011) survey of public opinion in 23 countries (including 12 of those with operating nuclear reactors), found public opinion against nuclear plants had risen significantly since an earlier poll in 2005 (see also Butler et al. in this book: 140–141). In countries with nuclear reactors, 39% said that no further reactors should be built. A further 30% said that operating plants should also be closed as soon as possible. The company noted that in none of these countries were the supporters of nuclear expansion in the majority, and that

> the proportion opposing the building of new nuclear power stations has grown to near-unanimity in Germany (from 73% to 90%), but also increased significantly in Mexico (51% to 82%), Japan (76% to 84%), France (66% to 83%), and Russia (from 61% to 80%). In contrast, while still a minority view, support for building new nuclear plants has grown in the UK (from 33% to 37%), is stable in the USA (40% to 39%), and is also high in China (42%) and Pakistan (39%).

A survey of 24 countries by *Ipsos-Mori* (June 2011) found similar regional variations, but three in five (60%) of all surveyed opposed the use of nuclear power. By contrast, there was high support for other forms of energy generation including solar power (97%), wind power (93%), hydroelectric power (91%), and natural gas (80%) as sources of electricity. A quarter of those surveyed said they had been influenced by the events at Fukushima.

Trust is one side of social quality which Weber (1922: 122) called *legitimacy*. For Weber, a social order possesses a high degree of legitimacy 'when it enjoys the prestige of being binding' (Weber 1964: 124–125). Over several decades leading up to the Fukushima crisis the nuclear industry made the claim that associated risks were a lesser evil than those of climate change, and that nuclear power was *essential* to make adequate reductions to greenhouse gas emissions. The initial signs after the Fukushima crisis seem to be that in conjunction with a lessening trust in the industry, the credibility of this claim is diminishing (again, also Hara in this book). Once lost, trust is doubly hard to regain, and the high scale of the industry is likely to contribute to its difficulties in rebuilding that trust.

CONCLUSION

This chapter began with one of this book's key questions: Should we continue with nuclear power as a key technology for a low carbon future? As already noted, in the end it is governments who must answer that question, and they will frequently go about examining the potential of nuclear power in terms of its capacity to maintain or increase energy security. Yet, as we have noted, whilst the choices may appear as simply technical or economic, these factors and their relative importance in defining 'security' will be significantly shaped by the larger context of the sociotechnical system in which the options are understood.

The similarities with the *Titanic* case remind us that the nuclear industry fits a class of systems referred to as high scale and high consequence vulnerability. In these, the scale is large not only in physical terms but also in terms of social organization, economic investment, and technical sophistication and complexity. Understood from the perspectives of energy development within a sociotechnical system, the goal of energy security may well mean different things to different communities, as it is shaped by their experience and evolving understanding of their social as well as technical needs.

The Fukushima crisis occurred when the nuclear industry had already experienced a long run of economic difficulties, whose multiple causes included inability to develop sufficient trust over a very long history. At the very time that the industry was making some headway in building legitimacy, at least for its potential role in reducing greenhouse gas emissions, it must now face the implications of Fukushima. The impacts will be felt along all the dimensions of the sociotechnical system.

In particular, an important impact of Fukushima is to further undermine trust in the nuclear power industry, at least for now, and to reduce its legitimacy especially in those populations who see an inequity existing between risks and benefits. At the same time, the lower scale of, and potentially closer relations between, renewable energy technologies, and their industries and users may more easily develop the trust necessary to support the development of these alternatives.

As a consequence the role of nuclear power in energy security is already more sharply contested in many countries and seems likely to be increasingly contested and undermined. Whether, when finally viewed in hindsight, this moment will be judged to have been a major tipping point for the industry remains to be seen. At a minimum, Fukushima Daiichi in all its aspects, will play a strong role not only in future demands of what constitutes safety in nuclear power, but more broadly in the debate about whether, as an institution, the nuclear industry is to be sufficiently trusted. The outcome of that debate is not yet clear. It remains to be established out of the continuing and now sharpened social and technical contest over how energy security is to be understood and achieved.

REFERENCES

Beyea, J. "The Scientific Jigsaw Puzzle: Fitting the Pieces of the Low-Level Radiation Debate." *Bulletin of the Atomic Scientists* 68 (2012): 13–28.

Bijker, W., T. Hughes, and T. Pinch (eds). *The Social Construction of Technological Systems: New Directions in the Sociology and History of Technology*. Cambridge, MA: MIT Press, 1987.

Brown, D. *The Last Log of the Titanic*. Camden, ME: McGraw-Hill, 2001.

Buongiorno, J., R. Ballinger, M. Driscoll, B. Forget, C. Forsberg, M. Golay, M. Kazimi, N. Todreas, and J. Yanch. *Technical Lessons Learned from the Fukushima-Daichii Accident and Possible Corrective Actions for the Nuclear Industry: An Initial Evaluation*. Cambridge, MA: MIT, May 2011.

Bupp, I., J. Derian, M. Donsimoni, and R. Treitel. "The Economics of Nuclear Power." *Technology Review*, February 1975, 14–25.

Camilleri, J. *The State and Nuclear Power: Conflict and Control in the Western World*. Australia: Pelican, 1984.

Department of Resources, Energy and Tourism. *National Energy Security Assessment*. Canberra: Australian Government, 2009.
Dickens, C. *A Tale of Two Cities*. With Illustrations by H.K. Browne. 1st ed. London: Chapman and Hall, 1859.
Edwards, R. "Revealed: British Government's Plan to Play Down Fukushima." *Guardian*, June 30, 2011, accessed August 19, 2011. http://www.guardian.co.uk/environment/2011/jun/30/british-government-plan-play-down-fukushima.
Elliston, B., M. Diesendorf, and I. McGill. "Simulations of Scenarios with 100% Renewable Electricity in the Australian National Electricity Market." *Energy Policy,* 2012. doi:10.1016/j.enpol.2012.03.011.
Falk, J. *Global Fission: The Battle over Nuclear Power*. Melbourne: Oxford University Press, 1982.
Falk, J. "Fukushima Fall-out." *Arena Magazine* 112 (July 2011): 18–23, reprinted as "Nuclear Power after Fukushima: The Japan Crisis Has Raised the Stakes for Global Nuclear Policy." *Arena Features,* November 2011, accessed June 7, 2012. http://www.arena.org.au/2011/11/nuclear-power-after-fukushima/.
Falk, J. "Rethinking Energy Security in a Time of Transition." In *Energy Security in the Era of Climate Change: The Asia Pacific Experience*, edited by L. Anceschi and J. Symons, 240–254. UK: Palgrave MacMillan, 2012.
Garzke, W. Jr., D. Brown, and A. Sandiford. "The Structural Failure of the Titanic." *Oceans* 1 (1994): 138–148.
Garzke, W. Jr., T. Foecke, P. Matthias, and D. Wood. "A Marine Forensic Analysis of the RMS Titanic." *Oceans 2000*. Washington, DC: MTS/IEEE, 2000: 673–690.
GlobeScan. "Opposition to Nuclear Energy Grows: Global Poll, London UK." GlobeScan International, 2011, accessed May 27, 2012. http://www.globescan.com/commentary-and-analysis/press-releases/press-releases-2011/94-press-releases-2011/127-opposition-to-nuclear-energy-grows-global-poll.html.
Grier, P. "Was Chernobyl Really Worse than Fukushima?" *Christian Science Monitor,* April 26, 2011, accessed May 27, 2012. http://www.csmonitor.com/USA/2011/0426/Was-Chernobyl-really-worse-than-Fukushima.
Hall, W. "Social Class and Survival on the S.S. Titanic." *Social Science and Medicine* 6, 22 (1986): 687–90.
Hongo, J. "NGO Finds High Levels in Safe Area." *Japan Times*, March 31, 2011, accessed June 4, 2012. http://www.japantimes.co.jp/text/nn20110331a2.html.
Howells, R. *The Myth of the Titanic*. London: Macmillan, 1999.
International Energy Agency. *World Energy Outlook, Executive Summary*. France: OECD/IEA, 2011.
International Energy Agency. "Energy Security." 2012a, accessed May 26, 2012. http://www.iea.org/topics/energysecurity/
International Energy Agency. "Energy Technology Perspectives: Pathways to a Clean Energy System." 2012b, France: OECD/IEA.
International Panel on Climate Change (IPCC). *Special Report of the Intergovernmental Panel on Climate Change*. Cambridge: Cambridge University Press, 2012.
Ipsos-Mori. "Strong Global Opposition Towards Nuclear Power." *Ipsos-Mori,* June 23, 2011, accessed May 27, 2012. http://www.ipsos-mori.com/researchpublications/researcharchive/2817/Strong-global-opposition-towards-nuclear-power.aspx.
Japan Times. "Cabinet Kept Alarming Nuke Report Secret." *Japan Times,* January 22, 2012, accessed November 4, 2012. http://www.japantimes.co.jp/text/nn20120122a1.html.
Japan Times. "No-go Zone Trespassers Face Fines, Arrest." *Japan Times,* April 22, 2011: 1.
Kang, J. "Five Steps to Prevent Another Fukushima." *Bulletin of the Atomic Scientists*, May 4, 2011, accessed May 7, 2012. http://www.thebulletin.org/web-edition/features/five-steps-to-prevent-another-fukushima.

Karr, J. *Les Guêpes*. January 1849.
Komanoff, C. "Cost Escalation in France's Nuclear Reactors: A Statistical Examination." January 2010, accessed May 4, 2011. http://www.komanoff.net/nuclear_power/Cost_Escalation_in_France's_Nuclear_Reactors.pdf.
Koomey, J., and N. Hultman. "A Reactor-Level Analysis of Busbar Costs for US Nuclear Plants 1970–2005." *Energy Policy* 35 (2007): 5630–5642.
La Porte, T. (ed). *Social Responses to Large Technical Systems: Control or Anticipation*. Netherlands: Kluwer Academic Publishers, 1989.
Massachusetts Institute of Technology Study Group. *The Future of the Nuclear Fuel Cycle*. Cambridge, MA: MIT Press, 2011.
McCluskie, T., M. Sharpe, and L. Marriott. *Titanic & Her Sisters Olympic & Britannic*. London: PRC Publishing, 1998.
McKenzie, Debora. "Caesium Fallout from Fukushima Rivals Chernobyl." *New Scientist*, March 29, 2011, accessed May 7, 2012. http://www.newscientist.com/article/dn20305-caesium-fallout-from-fukushima-rivals-chernobyl.html.
McKnight, H., and N. Chervany. "The Meanings of Trust." Management Informations Systems Research Centre (MISRC), Technical Report 96–04, 1996.
Meshkati, N. "Human factors in Large-Scale Technological Systems' Accidents: Three Mile Island, Bhopal, Chernobyl." *Organization & Environment* 5 (January 1, 1991): 133–154.
Miller, C., D. Sarewitz, and A. Light. "Science, Technology, and Sustainability: Building a Research Agenda." Report, National Science Foundation, Arlington, Virginia, September 8–9, 2008.
Moe, E. "Vested Interests, Energy Efficiency and Renewables in Japan." *Energy Policy* 4 (2012): 260–273.
Mohrbach, L. "The Defence-in-Depth Safety Concept: Comparison between the Fukushima Daiichi Units and German Nuclear Power Units." *atw-International Journal for Nuclear Power* 56, nos. 4–5 (2011): 1–10.
Moriarty, P., and D. Honnery. "What Is the Global Potential for Renewable Energy?" *Renewable and Sustainable Energy Reviews* 16 (2012): 244–252.
NEI (Nuclear Energy Institute). "World Nuclear Power Generation and Capacity as of March 2012." March 2012, accessed June 9, 2012. http://www.nei.org/resourcesandstats/documentlibrary/reliableandaffordableenergy/graphicsandcharts/worldnucleargenerationandcapacity/.
New Scientist. "Japan's Last Operational Nuclear reactor to Go Offline." *New Scientist*, April 24, 2012, accessed May 7, 2012. http://www.newscientist.com/article/mg21428624.100-japans-last-operational-nuclear-reactor-to-go-offline.html?DCMP=OTC-rss&nsref=online-news.
O'Connor, P. "Death Sat upon the Bow of the Titanic." *Pilot*, April 13, 2012, accessed May 13, 2012. http://www.thebostonpilot.com/article.asp?ID=14561.
Osang, A. "In the Shadow of Fukushima: A Mayor's Battle to Keep His Displaced Town Together." *Spiegel Online,* April 14, 2011, accessed May 7, 2012. http://www.spiegel.de/international/world/0,1518,druck-756647,00.html.
Ottmers, R. "Titanic Inquiry Project: Electronic Copies of the Inquiries into the Disaster," n.d., accessed 6 June 2012. http://www.titanicinquiry.org/.
Perrow, C. "Fukushima, Risk and Probability: Expect the Unexpected." *Bulletin of the Atomic Scientists*, April 1, 2011, accessed April 12, 2011. http://www.thebulletin.org/web-edition/features/fukushima-risk-and-probability-expect-the-unexpected.
Powell, B., and H. Takayama. "Fukushima Daiichi: Inside the Debacle." *Fortune Tech*, April 20, 2011, accessed May 12, 2012. http://tech.fortune.cnn.com/2012/04/20/fukushima-daiichi/.
Sawa, A. "Conflicting Policies: Energy Security and Climate Change Policies in Japan." In *Energy Security in the Era of Climate Change,* edited by L. Ancheschi and J. Symons, 127–128. UK: Palgrave Macmillan, 2012.

Schneider, M., A. Froggatt, and S. Thomas. *The World Nuclear Industry Status Report 2010–2011: Nuclear Power in a Post-Fukushima World: 25 Years after the Chernobyl Accident.* Washington, DC: Worldwatch Institute, April 2011.

Scott, R. *The History of the International Energy Agency.* Volume 2, *Major Policies and Actions of the IEA*. Paris: OECD/IEA, 1994.

Segawa, M. "Fukushima Residents Seek Answers Amid Mixed Signals from Media, TEPCO and Government." *Asia-Pacific Journal,* May 16, 2011, accessed August 19, 2011. http:// japanfocus.org/-Makiko-Segawa/3516.

Smith, K., Jr., P. William, H. Garzke Jr., R. Dulin Jr., F. Bemis., and C. Filling. "Marine Forensics—Historic Shipwrecks Determination of the Root Cause." *Oceans* 1 (2005): 432–440.

Ten Hoeve, J., and M. Jacobson. "Worldwide Health Effects of the Fukushima Daiichi Nuclear Accident." *The Royal Society of Chemistry/Energy & Environmental Science* DOI: 10.1039/c2ee22019a, June 26, 2012.

Weber, M. *Gundriss der Sozialökonomik*. Tubingen: LCB Mohr, 1922.

Weber, M. *The Theory of Social and Economic Organization.* US: The Free Press, 1964.

Webster On-Line Dictionary, 2012, accessed May 30, 2012. http://www.definitions.net/definition/trust.

Yamaguchi, M. "Japan PM Wants Less Reliance on Nuclear Power." *Boston Globe*, July 13, 2011, accessed August 19, 2011. http://www.boston.com/business/articles/2011/07/13/japan_pm_wants_less_reliance_on_nuclear_power/.

10 The Future Is Not Nuclear

Ethical Choices for Energy after Fukushima

Andrew Blowers

TURNING POINT OR BLIP?

By any measure—technical, social, political—the accident that overwhelmed the Fukushima Daichii nuclear power station (otherwise known as 'plant') in March 2011 was a catastrophe, not only for Japan but for the world. In terms of scale, impacts and implications the disaster ranks alongside that of another nuclear catastrophe a generation earlier in 1986 on the other side of the world at Chernobyl in Ukraine. In both cases the nuclear industry seemed to stop dead in its tracks. Chernobyl put the nail in the coffin of an industry already in retreat; Fukushima appeared to stall the promise of a 'nuclear renaissance' just gathering speed as countries sought solutions to problems of electricity supply and global warming. An increasingly confident nuclear industry was proclaiming the necessity, if not the inevitability, of a nuclear future when the tsunami struck. Yet, neither Chernobyl nor, it seems, Fukushima have yet dealt the knockout blow to nuclear's ambitions and pretensions as the salvation to the energy problem.

The political response has been uneven with many countries pondering their nuclear future, some backing off and others forging forward. All have had to confront the question of whether to continue with nuclear energy or not; a question that also constitutes a key objective of this book about the implications of the Fukushima Daiichi disaster. The question has been framed in a number of ways: economic—is it affordable? technical—is it safe? political—is it acceptable? These are the practical, pragmatic and policy issues about nuclear's part in the energy mix. But, Fukushima is a moral issue, too; it reignites the enduring and overarching question, is it right? This is a question that strikes at the heart of the morality of nuclear energy itself. It provokes the idea that there should be socially imposed limits to the application of complex technologies that potentially can release catastrophic consequences to environments and life. It is a question that invites us to consider some ethical issues about our relationship to future human society and to the future of the planet itself.

What constitutes the ethical context of the choice between nuclear or not? Ethics act as a guide to what is acceptable or unacceptable, what we should do, what is right or wrong, good or bad. 'Ethics are about how we *ought* to act, in contexts that have significant implications for human and non-human lives and well-being' (Rawles 2007: 26). Of course, ethical concerns are only part of the context for decision making and need to be considered alongside other circumstantial issues that will determine the future of nuclear energy (as addressed in many other chapters in this book). Nonetheless, it may be said that ethical concerns require a focus on the fundamental issue, whether to proceed or withdraw, whether to be or not to be? This existential ethical question can be posed in more obviously empirical terms as: *Are there any circumstances in which it would be acceptable to continue with nuclear as an energy option?*

This question is implicit in the debates about the future of nuclear energy post-Fukushima Daiichi though, in Germany at least, it has become a more explicit component of policy making (Ethics Commission 2011; see also Butler et al., and Falk, in this book). The choice is not simply between nuclear or not (Elliott 2007), but, for practical policy and political reasons, there are variants. At one end there is the option of complete and immediate shut down either temporary or permanent; at the other end, the decision to continue uninterrupted. In practice neither of these extremes has occurred except in Japan itself where, for a time, all of its post-Fukushima remaining reactors (50) were shut down for maintenance or safety reassessment. For other nuclear countries the choice lay between two sets of alternatives. One was to phase out or scale down existing nuclear programs and to abandon or reject any new nuclear programs. The other was to review nuclear safety in the light of the disaster and seek to ensure that technical and safety adjustments were made where possible to existing plants and applied to new designs.

In the first group are countries like Germany, Switzerland, Belgium and Taiwan opting for phaseout; Italy, following a referendum, overwhelmingly against the reintroduction of nuclear power; and, potentially, France where shifts in public and political opinion presages a reduction in that country's massive nuclear program from producing three-quarters to half the country's electricity. In the US where no new reactors have been ordered since 1978 but where a nuclear renaissance appeared to be gathering ground with potentially 28 new reactors, Fukushima Daiichi has effectively derailed the stuttering resurgence (Duffy 2011). In Japan, stringent conditions were imposed on reopening its reactors, and future nuclear policy was placed in the melting pot (IISS 2012).

But, for the greater part, countries have opted to carry on. This is especially so in the Far East where China, with the world's largest new build program, is likely to carry it forward having paused for inspection of all plants. India, too, has undertaken safety and regulatory reviews. In Europe, reactors are under construction only in France and Finland while the UK has the most ambitious new nuclear program calling for between 10 and 14 GW (gigawatts) of electricity from power stations at eight possible sites (DECC 2011) though, for various reasons, this is more likely to turn out to be fantasy rather than fulfillment. Overall, and so far at least, it seems the Fukushima effect has resulted in a slowing down of nuclear programs, a time to pause and reflect before opting either for phase out or business as usual.

The unevenness of the response is, in part, a pragmatic reaction to perceptions of public anxiety. Up to Fukushima Daiichi there had been a noticeable revival in public support for nuclear programs. For instance, by 2009 around 59% of the American public favored nuclear power (Hale 2012: 264). A poll for the Energy Security Initiative at Brookings taken shortly after the disaster indicated a roughly even split between those (46%) who felt nuclear power is necessary and those (48%) who considered the dangers too great. In Europe, the Eurobarometer poll traced a gradual improvement in nuclear's showing during the critical period of 2005 to 2008 when support for nuclear energy production rose by 7% to 44% totally or fairly in favor while opposition dropped to 45%, a decline of 10% of those totally or fairly opposed (Eurobarometer 2008).

After Fukushima Daiichi, that public support dropped markedly with a poll for the BBC in 23 countries indicating 22% in favor of expanding nuclear, 71% believing it could be replaced by alternatives, 39% in favor of using existing reactors and 30% opting for total shut down. Opposition to nuclear new build was strong among eight countries surveyed, with Germany 90% against, France 83%, Russia 83% and Japan 84%. Only the UK bucked the downward trend with support for new nuclear rising from 33% in 2005 to 37% in 2011 while the US remained stable in support for new nuclear at around 40%, similar to China and Pakistan (BBC 2011). Clearly, the polls need careful interpretation since the samples, questions and assumptions vary but, in the main, they reveal a Fukushima factor in the dip in nuclear's support like the dip that occurred in the aftermath of Chernobyl (see also Butler et al. in this book: 141). It is, of course, far too early to say whether Fukushima represents a final turning point or merely a blip on an upward trend that appeared to have set in as nuclear was promoted as a safe and low carbon solution to the climate change issue. This takes us back to the question of whether to continue with nuclear energy or not.

A NUCLEAR ACCIDENT ANYWHERE IS A NUCLEAR ACCIDENT EVERYWHERE

Consider the scale, severity and consequences of the accident that befell the Fukushima Daiichi nuclear plant on March 11, 2011. The accident ranked 7, the highest on the International Atomic Energy Agency (IAEA) scale, leading to loss of cooling, partial meltdown of three reactors in the six reactor complex and the exposure of spent fuel stores leading to massive radiation releases to the atmosphere and ocean. The total release is contested (Nature 2011), with varying estimates of potential deaths and other health effects in the future. It will be many years, if ever, before the total release and its health consequences are known. What is already clear are the long-term economic and psychological effects that will be caused by the evacuation of the zone around the plant with up to 100,000 people unlikely to return to their homes and livelihoods in the foreseeable future. The impact on the nuclear industry in Japan was devastating, a palpable demonstration that a major accident at one location cascades to shut down plants elsewhere both in Japan and around the globe. Thus, the biggest threat to the future of the nuclear industry is the nuclear industry itself. A nuclear accident anywhere is a nuclear accident everywhere.

By any reckoning, the Fukushima Daiichi disaster was a cataclysmic event, creating widespread ecological devastation, displacement of population, economic catastrophe, social disruption and psychological trauma. And it is clear that these impacts will persist, many of them into the far future. These are the circumstances that contextualize the moral issue. What is now at issue is at what scale and in what circumstances would a future routine, accidental or deliberate release of radioactivity be acceptable? Is what happened at Fukushima sufficiently terrible for a repetition to be regarded as unthinkable? Or, traumatic as it was, should Fukushima be regarded as a rare event and the remote possibility of a major accident a risk worth taking for the benefit of having nuclear energy?

While an ethical analysis will not yield necessarily definitive answers to these questions, it does provide a basis for making reasoned judgments based on our own perspectives. Ethical reasoning should enable us to reevaluate why we should or should not continue with nuclear energy in the light of Fukushima Daiichi. Ethical considerations enable values to be applied to policy options in order to justify choices made. Hence, ethical justification may be appealed to by both those who favor nuclear energy and those who seek its abandonment. In what follows I look at these alternative positions and the ethical arguments deployed to justify them. I shall address two claims made by the pronuclear camp. First is their claim that such an accident cannot happen again. Second, is that, put in context, Fukushima is far less catastrophic than the consequences of climate change that might be mitigated by the large scale deployment of nuclear energy. I shall then go on to argue that nuclear's putative contribution to carbon reduction is likely to be very small and, in any case, not needed when safer low carbon and renewable options are available (see also Lowe in this book: xxvi). That being so there is no ethical case for an activity that is associated with large risks from accidents, nuclear weapons and highly radioactive wastes that will persist into the far future.

IT WILL NOT HAPPEN AGAIN

In the past, denial or dismissal was the first response of pro-nuclear interests in the face of accidental releases of radioactivity. There is a history of early cover-ups such as Windscale in the UK in 1957 (Breach 1978). At around that time the biggest cover-up of all occurred when a level 6 accident involving radioactive waste scattered over 1,000 square kilometers in the Urals in the former Soviet Union leading to restricted access and affecting thousands of people. It was not revealed until two decades later (Medvedev 1979). In the US, cover-up and scaling down the number and potential consequences was routinely practiced by the government–nuclear industry complex (Shrader-Frechette 2011). And there have been over the years many other 'incidents' (unforeseen events and equipment failures mostly contained within sites), which have received little publicity. Sovacool (2011) records 76 substantial accidents involving either loss of life or more than US$50,000 worth of damage during the half century 1947–2008. By the time of Three Mile Island and, more significantly, Chernobyl, it was no longer possible to cover up the scale of devastation that could be caused by a major accident. Denial was not an option, and denial can only be described as unethical.

A more nuanced and ethically justified pronuclear position is to argue that Fukushima Daiichi was a unique, one-off, unrepeatable event. 'It could not happen here' is a response that seeks to justify continuing with nuclear energy on the grounds that such accidents are 'black swan events', extremely rare, unpredictable and avoidable only provided that measures are taken to ensure they are not repeated (Möller and Wikman-Svahn 2011). This was a near universal reaction among governments and nuclear operators in the wake of Fukushima. Everywhere, reactors were put under inspection to ensure that the failures encountered at Fukushima Daiichi would not be repeated. Thus, checks to ensure the viability and integrity of earthquake resistance, flood protection and power supply would be proof against a similar combination of the circumstances that had overwhelmed Fukushima Daiichi. In some cases, reactors were regarded as too vulnerable and were shut down. Germany was the extreme case, cancelling planned life extensions, shutting eight reactors permanently with the remaining reactors to shut down by 2022. China, with the biggest nuclear program, undertook safety assessments and temporarily halted construction at 26 new reactors for inspections.

In the UK, a major review of the country's reactors by the Chief Inspector of Nuclear Installations was conducted 'to examine the circumstances of the Fukushima accident to see what lessons could be learned to enhance the safety of the UK nuclear industry' (ONR 2011a, Executive Summary). In his reports the Chief Inspector made a range of recommendations, notably on flooding, seismic resilience, reactor integrity, spent fuel management, emergency response and planning, all of which would need to be followed up. But, once implemented, the report considered the plants could be rendered safe against even extreme natural hazards such as had combined to imperil Fukushima. Crucially, Weightman concluded: 'There is no need to change the present siting strategies for new nuclear power stations in the UK' (ONR 2011a, Executive Summary: 68). 'Flooding risks are unlikely to prevent construction of new nuclear power stations at potential development sites in the UK over the next few years' (ONR 2011b: 153). For the UK, it seems, a Fukushima could be prevented without impediment to the nuclear program.

A similar position was declared for Europe as a whole. The European Commission (EC) established a regime of 'stress tests' covering 147 nuclear reactors in 15 European Union (EU) countries together with 15 reactors in Ukraine and five in Switzerland. The work was 'of exceptional nature from a quantitative and qualitative point of view' (ENSREG 2012). The review and its 17 country reports had two parallel tracks, one concerned with safety issues arising from external natural events as at Fukushima Daiichi, the other a security review analyzing threats from terrorism. The outcome of this process was to stress the need for periodic reviews, to implement measures to protect containment integrity and to focus on measures to prevent accidents but also to limit their consequences. Nonetheless, it was left to individual countries to decide how and when to implement any measures and the stress tests were more in the nature of ensuring a regulatory ethos of 'continuous improvement' than an attempt to ensure immediate remedy. The report for the UK was broadly content with the safety of the country's reactors and saw no reason for curtailing the operation of nuclear power plants or other nuclear facilities. As to siting, the Weightman report concluded: 'There is no need to change the present siting strategies for new nuclear reactors in the UK' (ONR 2011c: 170).

IT COULD NOT HAPPEN HERE

Although these reports did not explicitly rule out future catastrophic accidents, the whole tenor of them is that such events are preventable through prudent responses and management. The emphasis was on the improbability of accidents, and, by implication, the ethical argument against nuclear is relatively weakened. If the probability of an accident can be rendered vanishingly small then the high consequences of risk are tolerable; the nuclear imperative triumphs. Yet, is the risk so improbable as to enable a justification for continuation to be morally acceptable? While a combination of an earthquake measuring almost 9 on the Richter scale followed by a 15-meter wall of water taking out four reactors is virtually impossible (though, remember, it did happen), comparable events cannot be ruled out. A high proportion of nuclear plants are sited on major rivers or low lying coasts, the latter potentially vulnerable to inundation from impacts of sea level rise and storm surges, likely to increase in the coming years as the impacts of climate change become more evident. Certainly the eight sites selected for new nuclear stations in England and Wales (Scotland has taken a no new nuclear position) are all sited on coasts and estuaries (Blowers 2010).

Earthquakes, flooding, storm surges, tsunamis and coastal processes are not the only potential causes of a failure leading to meltdown. Human failures must also be taken into account leading to multiple and interactive causes. In the cases of Chernobyl and Third Mile Island, meltdown was brought about by a combination of causes including component failure, misunderstanding of what was happening, unanticipated events and incomprehensibility on the part of operators. While Fukushima, or the other catastrophes, will never be exactly replicated, it is quite conceivable that another major accident will occur at some time, somewhere, brought about by a combination of unforeseen (and unforeseeable) circumstances. Rather than being unique and improbable, the complex technology of nuclear makes accidents repeatable and quite possible. They occur far more frequently than the US Nuclear Regulatory Commission's (NRC) prediction of one core-melt accident every 1,000 years in the US and one every 250 years worldwide (cited in Shrader-Frechette 2012: 268). There have been at least five core meltdowns in roughly 50 years in the US and 26 worldwide (Shrader-Frechette, 2012), hardly an insignificant total and, given problems of cover-up, classification and definition, the totals could well be higher still. From time to time such accidents may be catastrophic. As Charles Perrow (1999: 3–4) put it, in the wake of Third Mile Island:

> This is not good news for systems that have high catastrophic potential, such as nuclear power plants. . . . It suggests that the probability of a nuclear plant meltdown with dispersion of radioactive materials to the atmosphere is not one chance in a million a year, but more like one chance in the next decade.

This proved to be an uncannily accurate prediction since Chernobyl occurred within a decade of Third Mile Island, followed by Fukushima 25 years later, with many smaller incidents in between.

The reality seems to be that far from being almost inconceivable, nuclear accidents are what Perrow describes as 'normal'. They are normal accidents in the sense that they are the outcome of highly complex, interacting material and organizational systems. They involve what he calls the 'tight coupling' of interdependent components opening up the possibility of potential failures, virtually impossible to foresee and difficult to comprehend. Failures occur in the material components (such as cooling as at Fukushima) and in the organizational systems (monitoring, inspection, control, maintenance). With such system characteristics, 'multiple and unexpected failures are inevitable' (Perrow 1999: 5). Moreover, it may be speculated that the frequency and scale of nuclear accidents are likely to increase as the number of reactors worldwide grows, as the size of reactors increases, as new and untested designs are introduced and as more reactors are grouped together in multireactor sites (as was the case with the six reactors at Fukushima Daiichi). Sovacool (2010) considers the pressures to sustain power supply and make profits in a privatized system of production, which encourages plant to stay online as much as possible with possible compromise to safety. Even with the strong safety culture that pervades the nuclear industry no amount of lessons learned, precautionary principles or regulatory requirements, is likely to be proof against the myriad things that can go wrong in such complex, high tech, high risk systems.

The argument that it cannot happen here is buttressed by the flurry of inquiries seeking out cause, attributing blame and making preventative recommendations such as the stress tests following Fukushima. But, as Fukushima and Chernobyl and others before them have demonstrated, the claim that it (not necessarily the same or similar but a catastrophic accident) cannot happen is true only until it does. In the face of this proposition, reactions have been varied as we have seen. Fukushima Daiichi forced a reappraisal of the safety of nuclear programs already troubled by mounting costs. But, it is likely that ethical considerations played a part in the decisions whether to carry on with business as usual or to phase out or scale down. For some countries, Fukushima provided a decisive answer to the ethical question posed earlier—they, in effect, decided there were no circumstances for which a release of radioactivity would be acceptable. For most countries, however, Fukushima provoked a more nuanced reaction and further justification for continuing with nuclear programs. The nuclear industry, with more or less support from governments, recovering from the shock of Fukushima, resumed its soothing and tendentious message that nuclear was a necessary, reliable, safe and secure component of the energy mix that will avert impending environmental disaster. The pronuclear absolutist ethical position that it (Fukushima) could not happen here was reinforced by a second line of defense, the relativist argument that the consequences are not so great as to rule out nuclear altogether. All we can truthfully say is that we simply do not know if, when, or where an accident might occur.

DOES IT MATTER?

Even if the possibility of a major accident is conceded, those in favor of nuclear energy might claim that nuclear is morally justifiable despite the potential for meltdown. As van de Poel (2012: 288) puts it: 'Not only is the probability of a meltdown low, the historical evidence thus far, especially from the Fukushima and Chernobyl disasters, suggests that the consequences are very serious but perhaps not that catastrophic that they provide reason enough to ban any experiment with nuclear energy'. There is a tendency among supporters of nuclear energy to play down the risks, to cover up or to underestimate the consequences, to be selective with the evidence (for Fukushima, see e.g. Hara, Juraku, and Morita et al., in this book). Shrader-Frechette (2012: 268) also points to the routine tendency for governments and the nuclear industry to indulge in 'data-trimming, relative frequency, and inconsistency' when providing information about nuclear accidents or costs.

Of course, such dissembling, assertion and manipulation is not limited to the pronuclear cause and a dialogue of the deaf has characterized the nuclear debate. On the one side are those who claim, for instance, that the deaths and health consequences arising from nuclear accidents are far lower than those from other energy producing activities, notably coal mining. After all, they claim, no one died as a result of radiation dose after Fukushima Daiichi, though over time it may be expected that deaths will be attributed to the release of radioactivity. On the other side, those like Sovacool (2010: 109) counter, that, in terms of fatalities, 'nuclear power ranks as the second most fatal source of energy supply (after hydroelectric dams) and higher than oil, coal, and natural gas systems'. The health impacts of Chernobyl remain deeply controversial given all the problems of attributing cause in a multi-variable situation with both immediate and long-term effects. Given the range of assumptions, samples and predictions it is quite impossible to gain any reliable insight into numbers involved but it is very easy to diminish or exaggerate the figures to support a point of view. To illustrate the point, an oft cited report by the Chernobyl Forum comprising intergovernmental and governmental agencies indicated 4,000 excess cancer deaths resulting from the radiation (Nature 2006). At a later press conference, though, the Chernobyl Forum stated the figure had been misappropriated and was neither definitive, scientific or adopted by the UN. Greenpeace, at the other extreme, suggested 270,000 cancers arising from Chernobyl of which 93,000 would be fatal (Greenpeace 2006). The point is not to attempt to unravel the conflicts and contradictions in reporting the impacts of major nuclear disasters, rather to demonstrate how figures may be used to suggest or support a particular ethical viewpoint.

While the number of deaths might be disputed, the trauma and devastation wreaked by such disasters as Chernobyl and Fukushima are indisputable. The impact must also be measured in terms of economic and social disruption to thousands of people. The World Health Organization (WHO) speaks of 'loss of economic stability, and long term threats to health in current and possibly future generations . . . feelings of helplessness and lack of control over their future' (WHO 2008). Altogether a total of 341,000 people were evacuated as a result of Fukushima and a third of these remains unable to return home, their land contaminated, their produce stigmatized, their lives hopelessly and irrevocably disrupted. Such impacts as these are hardly small; they are not even relatively small. Rather, they amount to the sacrifice of large areas of land, people and nature. Both Chernobyl and Fukushima were accidents of regional impact but with some global dimensions.

What is clear in the case of a nuclear accident on this scale is that there is really no end to the catastrophe, of predicting the final extent of the damage or the geographical scale over which it extends. It becomes impossible to place social, spatial and temporal limits on the consequences and so 'the conclusion drawn is that, if adverse events are to be ruled out, nuclear technology must no longer be used' (Ethics Commission 2011).

WORSE MAY HAPPEN

A further ethical argument deployed by the pronuclear camp is that however terrible a nuclear disaster may be, it pales to insignificance when compared to the global catastrophe that is presaged by climate change. By framing nuclear energy in the context of climate change a seductive ethical argument can be deployed. The risks from nuclear energy, most notably from nuclear waste, are regarded as far less threatening than the risks from global warming. This is precisely the argument used by the UK government to justify its new nuclear strategy. In what is known as the 'Justification' process required under European legislation, the UK government concluded that 'the detriments likely to arise from not taking action on climate change by investing in low-carbon forms of energy such as nuclear raise more significant challenges for future generations than the detriments likely to arise from the management and disposal of radioactive waste' (DECC 2009a: 77). While there are evident dangers to health and the environment from an accidental or deliberate release through terrorist attack, the chances of such an unplanned event are dismissed as 'negligible in the UK' (DECC 2009a: 120). A similar ethical argument was put by the government in a white paper on energy when considering whether to create new nuclear waste. The risk of not allowing new nuclear power to play a role in meeting carbon emissions targets had to be set against the risks from nuclear energy and waste. The view was that 'the balance of ethical considerations does not require ruling out the option of new nuclear power' (DTI 2007: 200).

This ethical posturing can be revealed as sophistry on at least three counts, which I develop over this section and the following two. First, is the assumption that nuclear is a necessary part of the low carbon energy mix to ensure mitigation of climate change. Some argue that nuclear is not particularly low carbon once the carbon arising from the full cycle including energy intensive uranium mining, high energy inputs into plant construction, and other energy intensive operations are properly accounted for. Quite aside from that is the evidence that nuclear is not, in fact, a necessary option but one with high opportunity costs (as well as soaring real costs), which diverts attention and resources from plausible alternative options for meeting carbon targets. A range of studies has been produced to show that it would be possible to meet most, if not all, electricity demand from renewable energy sources by the middle of the century in the UK and elsewhere. The actual mix depends on what assumptions are made about targets, demand, supply, efficiency, technology, investment costs, lifestyles, inequalities and other variables all of which are constantly changing (e.g. HM Government 2009, Skea et al. 2010, SRU 2011). As time goes by these possibilities are becoming ever more obvious and the need for nuclear, especially new nuclear, becomes ever more difficult to justify either rationally or ethically.

The point may be made by comparing the UK and Germany. The UK's energy strategy provides for 59 GW of new electricity generating capacity by 2025 (out of a total of around 113 GW) of which 33 GW would come from renewables, leaving 26 GW remaining. According to the UK government's National Policy Statement for Nuclear Energy, nuclear 'should be free to contribute as much as possible' (DECC 2010, Annex A: 27). The most optimistic expectation might be 16 GW, with a more realistic figure nearer 6 GW. Even that begins to look unattainable by 2025 as nuclear companies baulk at the high up-front costs and long term risks to anticipated returns on investment. Already in the UK, renewables (notably wind, and photovoltaic) are, at last, beginning to move forward with new or undeveloped technologies (wave, tidal, biomass) coming into view for the longer term. Meanwhile, the putative 'energy gap' is likely to be plugged by gas and possibly life extensions to nuclear power which are lower carbon than the coal-fired supply they will replace. Hence, the 'strategy' is rapidly becoming a futuristic fantasy as, in a privatized and competitive market, nuclear is likely to be displaced in any event. For Germany the situation is much clearer, with parliamentary approval of a phase out of nuclear by 2022 based on the belief that a 100% renewable electricity system by midcentury is possible, safe and affordable (SRU 2011). Moreover, the German government appointed an Ethics Commission to set out an energy supply that 'dispenses with nuclear energy as quickly as possible and advances Germany's route towards sustainable development and new models of prosperity' (Ethics Commission 2011: 9). The Commission concluded that the country's nuclear energy component could be replaced without impairing energy supply, importing nuclear or increasing fossil fuels. Therefore, the argument that there is no alternative to nuclear energy has no ethical purchase at all and the inevitable conclusion is 'namely to end the use of nuclear power stations as quickly as the power they supply can be replaced by lower-risk sources of energy, based on ecological, economic and social acceptability' (Ethics Commission 2011: 14).

The UK and Germany represent different ethical positions. The UK government uses the claims that nuclear can help to save the planet and to prevent the lights going out while Germany believes it is possible to cope sustainably without nuclear as part of the equation. In a world where about 15% of electricity is supplied by nuclear power, the practicality of coping without it in the near future may prove difficult in some cases. But, it is also the case that nuclear promises more than it can deliver. In almost every country, nuclear programs routinely fail to deliver on time or within budget. For example, the Flamanville plant in northern France was declared to be two times over budget in taking twice as long to complete (2016 rather than 2012) as had been planned. Likewise the Olkiluoto plant in Finland has experienced delays owing to design and construction problems. There is already serious slippage in the UK's nuclear program as potential investors ponder the long-term investment risks. In circumstances where new nuclear is neither necessary nor practicable it follows it is also unacceptable.

A RISK IN ITS OWN RIGHT

A second reason why nuclear risk should not be compared to the risks from climate change is that they are not comparable—nuclear is an ethical risk in its own right. In spatial terms it is an unbounded risk, with local, regional and potentially global impacts. The first public acknowledgement of radioactive release from Chernobyl came when abnormally high levels of radiation traced to the plant were detected at Forsmark in Sweden some 600 miles away and subsequently elevated levels were recorded, but not always publicized, around the world. In temporal terms, nuclear is an infinite risk, some radionuclides having half-lives of hundreds of thousands, even millions of years. Long after nuclear reactors are shut down, radioactive wastes remain in stores with no ultimate solution for their long-term management beyond the concept of deep geological disposal. No repositories for high level wastes and spent fuel are yet in operation though several countries have plans in place. But, to claim, as the UK government does, that it 'is satisfied that effective arrangements will exist to manage and dispose of the waste that will be produced from new nuclear power stations' (DECC 2009b: 25), is more a way of justifying its nuclear program than a strategy founded on firm evidence of implementation. Even if repositories are eventually commissioned there remains the possibility that radionuclides will find their way into the accessible environment at some distant point and cause harm to an unsuspecting population and to the environment.

Unlike climate change where the impacts could conceivably be mitigated through concerted action, the effects on health and environment from a nuclear accident or far future release from a repository are, so far as we know now, likely to prove irreversible. Furthermore, the huge costs arising from the impacts of a major accident, as the Fukushima case shows, cannot be borne by the nuclear operators but must be borne by the public authorities wherever and whenever they occur. Above a certain level of financial risk, nuclear accidents are uninsurable and externalized. In that the costs could be incalculable, they represent an unknown and unethical burden. Unlike climate change where cost and burdens are diffused through society and the economy at large, the costs of a nuclear accident can be attributable to individual nuclear plants and facilities and the infinite liabilities arising become, in effect, subsidized by the public to enable privatized nuclear operators to compete in energy markets. These and other subsidies to the nuclear industry become the sine qua non of keeping the industry in being at public expense while misleading the public about the true costs. As Duffy (2011: 678) commented: 'The bottom line is that nuclear power is simply not competitive without huge government subsidies, and if these subsidies were withdrawn, all talk of a nuclear renaissance would immediately cease'. And yet, for those who favor nuclear, subsidies provide the essential support for its continuation as an acceptable energy option.

THE ULTIMATE DETERRENT

The third reason why nuclear power should be regarded as an incomparable ethical issue is its inevitable connection to nuclear weapons. Nuclear energy was, initially, developed as the civil arm of the development of nuclear fission for military purposes. The idea of the 'peaceful use of the atom' indicates its secondary and, at first, subsidiary intention. The first nuclear power stations were purpose built as part of the production process for the uranium and plutonium used in the atomic bomb. The hulks of these early reactors are scattered across the US and the former Soviet Union. In the UK, the power station at Calder Hall primarily built to produce plutonium was the first nuclear power station to produce electricity for the national grid in 1956. It is interesting to recall the hopes and ethical innocence invested in nuclear power as a clean, cheap and safe source of energy in those days. 'This was a period when great hopes were entertained by those advanced nations which could, or soon would, build civil atomic power stations, that nuclear-generated electricity would, despite high initial capital costs, beat fossil fuels both in terms of cost-per-unit *and* in the cleanliness stakes' (Hennessey 2006: 123). Although there have been efforts to divorce the military and civil components of nuclear fission power, the interconnections inevitably remain. In the US the military production and decommissioning complexes are confined to large and remote reservations such as Hanford, Oak Ridge, Savannah River and Idaho. Military nuclear wastes are managed quite separately. In the UK the segregation is functional and institutional rather than physical and geographical. In other countries the security and safeguards regimes may vary in their tightness and application. In any case, the dangers of diversion and proliferation persist and can only increase as more states seek or attain nuclear weapons capability or materials fall into the hands of terrorist organizations.

It would be disingenuous to claim that abandoning nuclear energy would lead to nuclear disarmament. As for nuclear weapons, the genie is out of the bottle and the best that might be hoped for is a reduction of weapons stockpiles and a limitation of proliferation. In this, the curtailment of nuclear energy may have a part to play. Those countries, such as Germany, Italy and Switzerland, that have eschewed nuclear weapons are also countries that can contemplate forgoing nuclear energy. Nuclear weapons states like the US, the UK, China, France, or India and Pakistan, let alone those seeking to develop the nuclear cycle, most notably Iran, may be less willing to surrender a civil nuclear industry. The connections in terms of research, shared infrastructures and security systems, as well as potentially waste management, achieve a complementarity if not compatibility between civil and nuclear power. However, the connections also have undeniably sinister implications. The destructive capability of nuclear weapons poses a threat of global proportions, one that if it escalated would be utterly catastrophic. Climate change threatens the world with incremental environmental disaster; nuclear weapons with instantaneous obliteration. Nuclear energy might conceivably offer some carbon relief but this is likely to be small. But, to achieve that would require a continuing expansion and dispersal of nuclear power. It would increase the risk of the terrible consequences of major nuclear accidents and the terminal threat of nuclear Armageddon.

'Responding to climate change is ultimately a moral choice' (Hillman 2011: 9). But, responding to climate change does not mean that nuclear energy must be included in a low carbon mix to replace current high carbon energy sources as a part of that choice, indeed, the use of nuclear energy is itself a quite separate moral choice. The case against nuclear is, as I have tried to show, morally persuasive. To the claim that Fukushima cannot happen again, the answer is that nuclear accidents are normal and will occur sometime, somewhere in the future as they have in the past. To the argument that nuclear is necessary as part of the energy mix to secure a low carbon future, the answer is that there are alternative energy pathways that are feasible, cost effective, sustainable and safe. And, to the argument that nuclear must be seen as a solution to climate change rather than a risk in its own right, the answer is that nuclear poses a different but, in its terms, no less plausible or potent threat to the environment and human health especially when the connection to nuclear weapons is acknowledged. The question that remains is this: If nuclear energy is so ethically irresponsible why is it that some nations continue to persist with it? The answer, I think, lies in the power which nuclear represents and confers on its protagonists.

A QUESTION OF POWER

The choice of nuclear or not can be understood in terms of alternative discourses of power. These discourses are ways of framing debate and mobilizing resources to support a strategic policy choice. Ethical perspectives may be used to justify, explain and endorse the course of action that is taken. The ability to implement rests in the power relationships that exist at a particular point in time and place. To take two examples used earlier, Germany and Britain, quite opposite approaches to energy policy have been taken, each reflecting (and supporting) different ethical positions. In the German case, ethical concerns about nuclear risks have been explicitly employed to help confirm a nonnuclear direction. In the UK (or, more strictly speaking, England and Wales) a pronuclear direction emanates from a more positive construction of the benefits that nuclear energy conveys in terms of energy and environmental security. But, in both cases, energy policy derives from a set of power relationships embedded within a process known as 'ecological modernization' to describe how environmental factors are incorporated into modern political economies (Hajer 1995, Mol 1995, Weale 1992).

Ecological modernization has four strands. The first is *technological* whereby production processes are designed to achieve environmental protection through pollution controls and green technology. The second is an *economic* strand, giving primacy in policy making to the free market through emphasis on profit, competition and deregulation. This links to the third, a *political* strand, sometimes called 'political modernization' with the state adopting a facilitative or enabling role encouraging partnership and cooperative relationships (Tatenhove et al. 2000). This approach is extended to the fourth, a *societal* strand, whereby environmental groups in civil society are generally incorporated in policy making through 'cooperative environmental governance' (Glasbergen 1998). Altogether ecological modernization represents a basically optimistic approach to environmental policy making with its promotion of technological fix, a market economy and essential emphasis on business as usual.

From a pronuclear perspective the nuclear industry seems to fit comfortably into this paradigm. It claims to be a low carbon technology; it is, in most countries, privatized within a competitive market sector; politically it is regulated by a facilitative state; and there has been a trend toward limited participation in policy making from civil society. All of these features have propitiated the nuclear renaissance and, in some countries, have sustained its continuing presence despite Fukushima. In these countries power relationships are likely to be exclusive with privileged access to government enabling the maintenance of relatively centralized and closed decision making. Opposing elements in civil society may be granted participation through consultation but are denied privileged access to the government–industrial nexus where power resides. The UK is a prime example of what Dryzek et al. (2003) term the 'actively exclusive' state. Civil society may be vociferous but tends to be co-opted through consultation. In terms of influence over policy, civil society may be listened to but, on the nuclear energy issue at least, it tends to be unheeded.

This need not always be the case. A number of countries, as we have seen, have turned away from nuclear energy post-Fukushima. The political opportunities for dissent can be exploited by vigorous protest and even confrontation, as is the case in Germany. A major reason for the German policy turn was the mass protests against nuclear energy. In other countries, too, nuclear renaissance has stalled or been still born as a result of public and political reaction to which governments have felt it necessary to respond. The antinuclear turn reflects a different and more pessimistic set of principles embedded in an altogether darker discourse of power which was identified and termed 'Risk Society' by Ulrich Beck (1992). Indeed, Beck's thesis was conceived with the nuclear industry after Chernobyl much in mind.

Risk Society is concerned with 'the challenges of the self-created possibility, hidden at first, then increasingly apparent, of the self-destruction of all life on this earth' (Beck 1995: 67). Like Perrow, Beck diagnoses the dangers from complex high technology created, controlled and supervised by experts. But these technologies may ultimately be beyond expert control (as Fukushima so tragically demonstrated). Individuals experience anxiety and feel powerless in the face of high-consequence, involuntary, invisible, incomprehensible and unavoidable risk. Fears of a major accident and its consequences are not easily subdued post-Fukushima. In the aftermath of the accident there was plenty of evidence of Beck's analysis—the extent and pervasive nature of the disaster; the fallibility of control systems and problems of managing the accident; the closed and secretive nature of nuclear decision making; the inadequacy and paucity of information released to the public; the failures in emergency response—all compounded in a lack of civic trust and confidence in the ability of the nuclear operators and government to handle the crisis effectively. Both in Japan and elsewhere, Fukushima gave impetus to the ethos that there are no circumstances in which it would be acceptable to continue with nuclear as an energy option.

The ethical response to Fukushima has both reflected and influenced prevailing power relationships in different countries. In some countries the nuclear industry has proved remarkably resilient, an exemplar of ecological modernization and a necessary component of an energy mix to promote economic growth and environmental protection. This speaks to an ethical discourse of progress, growth and modernization. Conversely, the dangers from major accidents, routine emissions, from nuclear waste and the possibilities of proliferation and terrorism support a discourse of risk and its consequences and a language of disaster and degradation.

THE FUTURE IS NOT NUCLEAR

Whether or not nuclear energy is phased out or expanded there will be a legacy of radioactive wastes to be dealt with long after reactors have shut down. It may seem ethically wrong to add further to the risks of accidents and the legacy of radioactive wastes by building more nuclear power stations. It may also be argued that it is unfair to impose this burden of risk on specific places and on future generations, especially in terms of intergenerational equity that is a central premise of the so-called imperative of sustainable development. Fukushima Daiichi has mainly focused attention on the dangers from operating reactors. It is remarkable how little attention has been paid to the risks faced in the far future when decaying facilities and dangerous nuclear wastes may well remain in increasingly vulnerable locations such as the coastal sites in England and Wales, or those already in Japan and neighboring Taiwan. Beyond the next 50 years or so, predictions of the impacts of rising sea levels and accompanying impacts of climate change such as storm surges and coastal processes become increasingly speculative. Over such a timespan it also becomes increasingly difficult to be assured of societal stability or institutional continuity. In effect, burdens of cost, effort and risk are being imposed on future generations whose circumstances, social and environmental, are unknowable.

The present generation's obligation to the future is thus a moral issue. It may be questioned how far this obligation should extend. Some ethicists might argue that the obligation is infinite and has to extend to the limit of impact of our actions (Adam 2006:14). This idea of continuing responsibility may seem unyielding and impractical, but it does directly address the 'organized irresponsibility' of 'manufactured uncertainty' that Beck refers to in risk society. Perhaps more pragmatic and realistic is a criterion of diminishing responsibility, recognizing that increasing uncertainty will make it impossible to protect for the indefinite future. 'On sufficiently long time scales, any statement at all about the impacts of current actions and about obligations to the future become meaningless' (NEA/OECD 2006: 21). Nevertheless, concern for the far future should encourage a precautionary approach that certainly would rule out any further contemplation of nuclear expansion. At least the existing legacy of wastes is known; any further nuclear development will extend the time scales over which the burden must be managed for very long but essentially unknowable future periods.

The tenor of my argument has been that Fukushima is not a one-off event that can be prevented in the future; rather, it is a demonstration of what, inevitably, will happen again in the future. The consequences of major nuclear accidents are widespread and devastating and cannot be justified in terms of protecting environments or energy security. Alternative forms of development and efficiency will provide sustainable energy. Nuclear energy should not be regarded as part of the solution to climate change. It constitutes a mega risk of potentially global implications. Nuclear risk is unbounded yet uneven in time and space. It transgresses the criterion of sustainability as proposed by the IAEA that future society should be protected, 'in such a way that the needs and aspirations of the present generation are met without compromising the ability of the future generations to meet their needs and aspirations' (IAEA 1997, article 1).

Taken together, the prospect of further accidents and the long timescales of radioactive wastes refine and transform the overarching question posed at the outset of this chapter—Are there any circumstances in which it is acceptable for the present generation to impose a burden of nuclear risk on future generations? The ethical answer would seem to be that of all the technological options that should be considered for the future, nuclear energy is not one of them.

REFERENCES

Adam, B. "Reflections on Our Responsibility to and for the Future." In *Ethics and Decision Making for Radioactive Waste,* edited by A. Blowers, 12–19, London: Committee on Radioactive Waste Management (CoRWM), 2006.

BBC. "Nuclear Power Gets Little Support World Wide." *BBC News Science and Environment* 25 (November 2011).

Beck, U. *Risk Society: Towards a New Modernity.* London: Sage, 1992.

Beck, U. *Ecological Politics in an Age of Risk.* Cambridge: Polity Press, 1995.

Blowers, A. "Why Dump on Us? Power, Pragmatism and the Periphery in the Siting of New Nuclear Reactors in the UK." *Journal of Integrative Environmental Sciences* 7 no. 3, September (2010): 157–173.

Breach, I. *Windscale Fallout.* Harmondsworth: Penguin Books, 1978.

DECC (Department of Energy and Climate Change). *The Justification of Practices Involving Ionising Radiation Regulations 2004—Consultation on the Secretary of State's Proposed Decisions as Justifying Authority on the Regulatory Justification of the New Nuclear Power Station Designs Currently Known as the AP1000 and the EPR*. Volumes 1 and 2. London: DECC, 2009a.

DECC (Department of Energy and Climate Change). *Draft National Policy Statement for Nuclear Power Generation (EN-6)*. London: TSO, November 2009b.

DECC (Department of Energy and Climate Change). *Revised Draft National Policy Statement for Nuclear Power Generation (EN-6)*. Volume 2, Annex A. London: TSO, October 2010.

DECC (Department of Energy and Climate Change). *Overarching National Policy Statement for Energy (EN-1)*. London: DECC, URN 11D/711, 2011.

Dryzek, J., D. Downes, C. Hunold, D. Schlosberg, and H-K. Hernes. *Green States and Social Movements, Environmentalism in the United States, United Kingdom, Germany and Norway*. Oxford: Oxford University Press, 2003.

DTI (Department of Trade and Energy). *Meeting the Energy Challenge: A White Paper on Energy*. CM 7124, London: TSO, May 2007.

Duffy, R. "*Déjà vu* All Over Again: Climate Change and the Prospects for a Nuclear Power Renaissance." *Environmental Politics* 20, no. 5 (2011): 668–686.

Elliott, D. *Nuclear or Not? Does Nuclear Power Have a Place in a Sustainable Energy Future?* London: Palgrave Macmillan, 2007.

ENSREG (European Nuclear Safety Regulators Group). "Stress Tests and Peer Review Process, Joint Statement of ENSREG and the European Commission, 2012, Brussels," accessed 26 April, 2012. http://www.ensreg.eu.

Ethics Commission, The. "Germany's Energy Turnaround—a Collective Effort for the Future." Ethics Commission on a Safe Energy Supply. Berlin, May 30, 2011.

Eurobarometer and the EC. "Attitudes towards Radioactive Waste." *Special Eurobarometer* 297 (June 2008).

Glasbergen, P. (ed). *Co-operative Environmental Governance—Public-Private Agreements as a Policy Strategy*. Dordrecht: Kluwer Academic Publishers, 1998.

Greenpeace. *The Chernobyl Catastrophe, Consequences on Human Life*. Amsterdam: Greenpeace, 2006.

Hajer, M. *The Politics of Environmental Discourse*. Oxford: Oxford University Press, 1995.

Hale, B. "Fukushima Daiichi, Normal Accidents, and Moral Responsibility: Ethical Questions about Nuclear Energy." *Ethics, Policy and Environment* 14, no. 3 (2012): 263–265.

Hennessy, P. *Having It So Good, Britain in the Fifties*. London: Penguin Books, 2006.

Hillman, M. "Climate Change: *Quo Vadis et Quis Custodiet?*" *Environmental Law and Management* 23 (2011): 30–34.

HM Government. *The UK Low Carbon Transition Plan*. London: TSO, 2009.

IAEA (International Atomic Energy Agency). "Joint Convention on the Safety of Spent Fuel Management and on the Safety of Radioactive Waste Management, Information Circular INF/546." Vienna: IAEA, 1997.

IISS (International Institute for Strategic Studies). "Strategic Comments, Fukushima: the Consequences." 18, Comment 9, 2012, accessed March, 2012. http://www.iiss.org/publications/strategic-comments/past-issues/volume-18-2012/march/.

Medvedev, Z. *Nuclear Disaster in the Urals*, London: Angus and Robertson, 1979.

Mol, A. *The Refinement of Production*. Utrecht: van Arkel, 1995.

Möller, N., and P. Wikman-Svahn. "Black Elephants and Black Swans of Nuclear Safety." *Ethics, Policy and Environment* 14, no. 3 (2011): 273–278.

Nature. "Special Report: Counting the dead." *Nature* 440 (April 20, 2006): 982–983.

Nature. "Fallout Forensics Hike Radiation Toll." *Nature* 478 (October 25, 2011): 435–436.

NEA (Nuclear Energy Agency) and OECD (Organization for Economic Cooperation and Development). *The Environmental and Ethical Basis of Geological Disposal.* Paris: NEA/OECD, 2006.

ONR (Office for Nuclear Regulation). *Japanese Earthquake and Tsunami: Implications for the UK Nuclear Industry, Interim Report.* HM Chief Inspector of Nuclear Installations. Bootle: ONR, May, 2011a.

ONR (Office for Nuclear Regulation). *Japanese Earthquake and Tsunami: Implications for the UK Nuclear Industry, Final Report.* HM Chief Inspector of Nuclear Installations. Bootle: ONR, September, 2011b.

ONR (Office for Nuclear Regulation). *European Council 'Stress Tests' for UK Nuclear Power Plants, National Final Report.* Bootle: ONR, December, 2011c.

Perrow, C. *Normal Accidents: Living with High-Risk Technologies.* Princeton, NJ: Princeton University Press, 1999 (first published 1984, New York, Basic Books).

Rawles, K. "Ethics and Ethical Deliberation." In *Ethics and Decision Making for Radioactive Waste,* edited by A. Blowers, London: Committee on Radioactive Waste Management (CoRWM), February, 2007.

Shrader-Frechette, K. "Fukushima, Flawed Epistemology, and Black-Swan Events." *Ethics, Policy and Environment* 14, no. 3 (2011): 267–272.

Skea, J., P. Ekins, and M. Winskel. *Energy 2050: Making the Transition to a Secure Low-Carbon Energy System.* London: Earthscan, 2010.

Sovacool, B. "Critically Weighing the Costs and Benefits of a Nuclear Renaissance." *Journal of Integrative Environmental Sciences* 7, no. 2 (2010): 105–123.

Sovacool, B. *Contesting the Future of Nuclear Power.* Singapore: World Scientific Publishing, 2011.

SRU (German Advisory Council on the Environment). *Pathways towards a 100% Renewable Electricity System, Special Report.* Berlin: SRU, October, 2011.

Tatenhove, J. van, B. Arts, and P. Leroy (eds). *Political Modernisation and the Environment: The Renewal of Environmental Policy Arrangements.* Dordrecht: Kluwer Academic Publishers, 2000.

van de Poel, I. "Nuclear Energy as a Social Experiment." *Ethics, Policy and Environment* 14, no. 3 (2012): 285–290.

Weale, A. *The New Politics of Pollution.* Manchester: Manchester University Press, 1992.

WHO (World Health Organization). "Health Effects of the Chernobyl Accident: An Overview." Fact Sheet No. 303, April 2008.

11 Nuclear Emergency Response

Atomic Priests or an International SWAT Team?

Sonja D. Schmid

March 11, 2011, started and ended with television footage of the disaster at the Fukushima Daiichi nuclear power plant. Over the following days, I watched in horror as one reactor building after another exploded. When images of the first helicopter dousing water appeared on the screen, I could not help but remember the plight of thousands of Soviet soldiers and civilians trying to contain another nuclear nightmare, some 25 years ago. Where were those people now who had dealt with the Chernobyl disaster? Could someone dispatch them to Japan and let them help fix things up? But with every explosion that shook the Japanese plant it became clearer: there was nobody—not in Japan, nor Russia, nor the United States—who had the relevant know-how, equipment, and strategy to handle a nuclear disaster. No international nuclear emergency response group exists today.[1]

The dramatic events at Fukushima Daiichi will not remain the last severe nuclear accident the planet will have to face (Shrader-Frechette 2011, see also Blowers in this book). This realization marks a major shift in our thinking about nuclear risk, away from accident prevention, and toward accident mitigation and more rigorous emergency preparedness. Based on this insight, and in response to the Fukushima Daiichi crisis, the United States and nations in Europe and Asia have adopted widely different strategies ranging from a renewed commitment to nuclear expansion to significant safety overhauls, to announcements of a complete nuclear phaseout in the near future (see e.g. Chan and Chen 2011, Hong 2011, Jahn and Korolczuk 2012, Jiang 2012, Thomas 2012, US Nuclear Regulatory Commission 2011, Zhou et al. 2011). Regardless of the specific national roadmaps, however, nuclear safety has returned to the international stage with a vengeance.

The purpose of this essay is to discuss a truly formidable task, the creation of an international nuclear emergency response team. This idea aligns with a key objective of this book in exploring the implications of Fukushima Daiichi: What lessons can be learned from the inadequacies of the existing emergency response, in order to better manage future nuclear disasters? The idea of an international nuclear emergency team immediately poses a series of questions: Who would staff, train, and finance an international nuclear emergency response group? Is it even possible to develop an emergency response strategy that accounts for the uniqueness of each nuclear accident? And if so, how might the significant differences in technical designs, styles of operation, regulatory structures, and cultures of communication be overcome, and 'standardized'? Most importantly: Who would determine what kind of knowledge would be necessary and relevant? Subject expertise along with clear lines of command and control, will likely not suffice in a nuclear disaster. Which improvisation skills will need to be harnessed to provide effective nuclear disaster relief? How would such knowledge be accessed, archived, and taught? Considering such questions is important, in part because responding to the Fukushima Daiichi disaster will require much more than retroactive technical safety adjustments and ill-defined 'stress tests' (European Nuclear Safety Regulators Group 2012, US Nuclear Regulatory Commission 2011). The handling of the disaster has revealed and reinforced latent distrust in nuclear industry experts and the government agencies charged with regulating nuclear safety (Nuclear Accident Independent Investigation Commission 2012). Addressing this distrust will be difficult, but absolutely critical, when identifying or creating a group or agency that is both capable of assembling the needed expertise for effective emergency response, and that also is accepted as legitimate by the broader public.

CONTEXT AND ARGUMENT

The threat of nuclear disasters has captured the popular imagination at least since the bombings of Hiroshima and Nagasaki (e.g. Lente 2012, Weart 1988). The German sociologist Ulrich Beck, for example, argued in his 1986 bestseller, *Risk Society,* that radiation was a paradigmatic risk of modernity: invisible, irreversible, and 'manufactured' by the same technological progress that has helped us control other risks (also Beck 1992). The few significant disasters in the history of the world's nuclear industry have been unique events and hardly comparable, but they can all be interpreted as consequences of complex, tightly coupled environmental, social, and technical systems (Perrow 1984, Pritchard 2012). 'Fukushima', as this nuclear disaster is increasingly referred to (analogous to 'Chernobyl', or 'Three Mile Island'), has become a focusing event (Kingdon 2003: 94–100) that may allow not only broad policy changes, but also a fundamental reassessment of the processes of engaging the public (see also Hindmarsh in this book: chapter 4).[2]

My argument here utilizes primarily scholarship in science, technology and society (STS) studies, and additionally draws on work in organization theory, disaster sociology, and sociocultural studies of risk. As this book demonstrates, STS analysis not only brings the much-needed interdisciplinary perspective to the table, which an emergency response requires; it also foregrounds questions of expertise, trust, and legitimacy, as well as public engagement as part of that response (Collins and Evans 2002, Guston and Sarewitz 2002, Jasanoff 1994, Kleinman 2000, Nelkin 1992, Sclove 2010, Szerszynski 1999,

Wynne 1996, 2010). Numerous case studies have documented that meaningfully engaging lay communities in decisions traditionally made by scientific and technical elites can enable greater vigilance and raise confidence about individual emergency preparedness. By drawing on different perspectives, such engagement may also provide the benefit of better emergency planning.

Research in STS also explores how institutions that govern modern high-risk technologies acquire, maintain, and lose public legitimacy. Fukushima occurred in one of the world's most advanced economies, in a country governed by scientifically trained, technologically savvy elites.[3] The specific kinds of highly specialized knowledge involved with operating nuclear reactors however may not be accessible to broad public debate to the same degree as, for example, evacuation policies. But in the interest of sustainable, socially legitimate solutions, arguably decisions about even the technical responses to disasters should not be left to scientists and engineers alone, whether they are based within the nuclear industry, a regulatory body, or a nongovernmental organization. STS scholars, in particular, may be (and in fact, have been) called upon to help understand how seemingly arcane, technical decisions are inevitably shaped by norms, values, and attitudes (e.g. Fujigaki and Tsukahara 2011, also Juraku in this book).

The real challenge of a disaster involving nuclear facilities, however, lies in how to handle the unexpected, unpredictable, utterly novel, and barely intelligible chain of events unfolding in real time. A nuclear emergency response group can no doubt benefit from improving community resilience and emergency preparedness but this group will unavoidably carry an elite character. Along with better rules and standardized, interoperable hardware, there is simply no way around the creation of highly specialized technical elites. And yet, technical expertise alone does not automatically provide the authority to decide what is best for 'the public good'. In this respect, tomorrow's nuclear emergency responders will have to differ from the 'atomic priests' proposed in the 1970s and 1980s (Sebeok 1984, Weinberg 1972). These atomic priests were thought to ensure the longevity of institutions managing nuclear materials, and to guard nuclear facilities by restricting access to 'the truth' to a self-selecting corps of experts. By contrast, one of the critical lessons of Fukushima was the significance of addressing the social context adequately (as well discussed in other chapters in this book). A future nuclear emergency response team will need to establish clear, transparent criteria for success—both to attain public legitimacy, and to determine the most effective emergency response.

In the following, after briefly reviewing the course of events at Fukushima, I outline the kinds of knowledge and expertise a nuclear emergency response group would ideally need to provide. I compare it to what exists today and discuss possible candidates. I conclude that existing organizations with subject expertise have negligible international authority and often have problematic rapport with the general public, and confirm the need for a well-coordinated and integrated sociotechnical approach.

COURSE OF EVENTS

On March 11, 2011, an earthquake off Japan's eastern coastline triggered a tsunami that flooded the Tōhoku region.[4] The operating reactors at the Fukushima Daiichi plant shut down automatically. However, a nuclear reactor still requires significant electricity to provide cooling for the residual heat produced by radioactive decay in the core, even after a controlled shutdown. The massive tsunami knocked out not only the plant's connection to the electricity grid; it also flooded the buildings that held the back-up diesel generators, rendering them useless. The only redundancy left to prevent a station blackout were batteries, which were designed to bridge the time emergency teams needed to reconnect the station to the grid. But the tsunami's effect on the entire region was devastating. In addition to killing thousands of people, the tsunami cut off power lines; destroyed administrative buildings, access roads, and residential areas; and left huge areas cluttered with a thick layer of debris. It quickly became clear that more time would be needed to restore power to Fukushima Daiichi than the small window of opportunity that the station's batteries provided.

Over the ensuing days, one explosion after the other shook the plant and one reactor after another required emergency cooling. By then, only highly corrosive seawater was available, which ultimately could not prevent the reactors from overheating. There were concerns that the damaged cores might reach criticality anew (that is, a nuclear chain reaction would restart), and this threat would then spread from the reactors to the spent fuel pools, which were reportedly losing water (Wald and Berger 2011). Contaminated water was leaking—and later was deliberately released—into the ocean. All the while, vital infrastructure remained cut off and further earthquakes shook the region. Amidst the chaos, the Japanese prime minister stepped down in August, in response to intense criticism of his inept handling of the disaster (Takubo 2011).

Mismanagement was not the only charge mounted against the Japanese utility that operated the reactors at Fukushima Daiichi, Tokyo Electric Power Company (TEPCO). In the aftermath of the disaster, international media charged workers at the plant, alternatingly, with a lack of expertise to handle the situation adequately, and with a lack of courage, when they retreated temporarily under the threat of dangerously high radiation levels. Especially in the United States, experts criticized the design of Fukushima Daiichi's reactors (Bosk 2011, Zeller 2011).[5] And finally, TEPCO's record of falsifying reports in the past was used to attack the 'cozy relationship' between the Japanese nuclear industry and the regulatory agency (Aldrich 2011, Schmid 2011a, 2011b; see also Hara, Juraku, and Morita et al., in this book). Several months into the nuclear disaster, TEPCO finally declared that it was abandoning plans to return Fukushima Daiichi to operable conditions. The firm is now preparing to entomb the destroyed reactors after being saved from bankruptcy with government support (e.g. Butler et al.

2011, Morita 2012). And yet, TEPCO had elaborate emergency plans on the books for tsunamis (Government of Japan 2011). According to a report released in November 2011 by the US Institute of Nuclear Power Organizations (INPO), TEPCO personnel handled the extraordinarily difficult situation with remarkable composure and creativity (INPO 2011). This stands in marked contrast to other independent reports and broad public critique (e.g. Investigation Committee 2011, Nuclear Accident Independent Investigation Commission 2012; also Hara, and Morita et al., in this book).

But emergency preparedness is hardly ever considered 'good enough' in retrospect, especially after a disaster in which so many lives were lost or shattered. Given the wide variety of different perspectives, what are the criteria for successful emergency preparedness? Is the best we can hope for 'getting it right by accident'? There were good reasons (technical, economic, and managerial) for the way the Japanese nuclear industry was set up. A convincing case can be made for the siting of nuclear plants near the ocean or along rivers—even if this increases the seismic or flooding risk to these facilities. It also makes good economic sense to cluster several reactors on one site, rather than to create entire industrial infrastructures for a single reactor (cf. e.g. Juraku, and Hindmarsh, in this book: chapter 4). The right balance between safety and profitability remains elusive, as we know from other high-risk industries (Perrow 1984, 2007, Roberts 1990, Vaughan 2005). For all its undeniable flaws, the nuclear industry *worked* for several decades—in Japan and elsewhere. That is also the truly frightening realization after Fukushima: this disaster was not 'waiting to happen', but occurred in a system that had been functioning reasonably well for quite some time (Schmid 2011c, Vaughan 1996, cf. e.g. Hara, Juraku, and Falk, in this book).

REGIONAL AND INTERNATIONAL REPERCUSSIONS

Over the past decade, growing economies all over Asia had been developing nuclear energy aggressively, while European and North American nuclear industries were at best stagnating. The shock over Fukushima in the region was palpable.[6] In the light of mounting public concerns, expert critique, and the economic implications of Japan's nuclear disaster, national governments in the region reviewed their energy policies and in some cases revised their nuclear plans (Chan and Chen 2011, Hong 2011, Jahn and Korolczuk 2011, Jiang 2011, Zhou et al. 2011). Such was the scale and intensity of the disaster that international repercussions ensued: Germany announced its nuclear phase-out in late May 2012, followed soon after by Switzerland. In June 2012, the European Union began to administer 'stress tests' at European nuclear plants, and all US reactors underwent thorough inspections (US Nuclear Regulatory Commission 2011, European Nuclear Safety Regulators Group 2012).

The distress—both in the region and internationally—over the nuclear dimension of the March 11 disaster suggests that Fukushima Daiichi did in fact come as a surprise (Shrader-Frechette 2011). In the wake of the Three

Mile Island accident of 1979, which despite its relatively minor releases sent shockwaves through the nuclear industry worldwide, reactor safety had received a renewed boost (Walker 2004). During the three decades since then the nuclear industry and national regulators had increased their emphasis on avoiding nuclear accidents (Walker 1992). Notably, this approach relied on the idea that nuclear power plants were 'hostages of each other'—where an accident at one would affect them all (Rees 1994). The Fukushima Daiichi disaster has prompted a shift from this 'zero risk mind-set' to one that emphasizes preparedness for a nuclear emergency (Suzuki 2011).

Organizing an industry and communicating with the public in a context that suggests risk avoidance, however, is very different from encouraging risk acceptance. More than 20 years ago, social scientists Harry Otway and Brian Wynne cautioned that accident *prevention* (safer designs, better operator training, etc.), but even more so *emergency planning*, faced significant economic and managerial hurdles (Otway and Wynne 1989). In high-risk industries such as the nuclear power industry, safety sometimes gets pitted against profitability; security concerns often suggest less rather than more transparency; and commercial secrets all too often hamper intercorporate, let alone international, cooperation (Sagan 1993, Vaughan 1996).

In May 2011, France, the world leader in nuclear electricity generation, initiated the first post-Fukushima discussion on international nuclear accident mitigation. At that meeting, industry representatives from around the world agreed that operating a nuclear industry comes with profound responsibilities.[7] The deputy chief of Russia's nuclear operator, Rosatom's Nikolai Spassky, suggested that international law should force countries operating nuclear plants to abide by international safety standards.[8] This proposal amounts to a recognition of the international character of the nuclear energy industry, but it remains unclear as to who would enforce such rules and how—as of this writing, no international agency has such powers.

The global dimension of nuclear threats is old news to both the nonproliferation community concerned with preventing nuclear war, and the antinuclear movement protesting the operation of nuclear power plants. Neither radioactive fallout from nuclear weapons tests, nor accidental releases from commercial nuclear facilities respect national or jurisdictional boundaries. Severe nuclear accidents may thus require international institutions to coordinate their mitigation (Beck 1992, Lupton 1999). As if to confirm this point, the International Atomic Energy Agency (IAEA) held the Ministerial Conference on Nuclear Safety in June 2011, which emphasized the 'transboundary' character of nuclear accidents and acknowledged the importance of improving emergency preparedness and response to nuclear accidents at regional, national, and international levels (IAEA 2011).

In the wake of Fukushima then, the international community has come to acknowledge the magnitude of risk and responsibility involved in developing and safely operating nuclear facilities. Guaranteeing the safe operation of a national nuclear industry is increasingly considered the entire world's 'business' (e.g. Meshkati 2011). Consequently, we see a trend where mitigating the consequences of a nuclear disaster is also increasingly being regarded as an international task.[9] In turning to this task, flaws in existing nuclear emergency response plans need to be discussed.

FLAWS IN THE EXISTING NUCLEAR EMERGENCY RESPONSE

How can we base disaster response 'on scientific knowledge and full transparency', as the IAEA would like to mandate (IAEA 2011)? What knowledge should nuclear safety be based upon, where the science is still contested? And how useful is the notion of transparency in a context where the operation of nuclear power plants is considered an 'inalienable right', as the text of probably today's single most important nuclear treaty states (IAEA 1970)? Nuclear specialists around the world are still discussing the existing emergency response organizations and the reasons they ultimately failed at Fukushima Daiichi (Government of Japan 2011, INPO 2011, Investigation Committee 2012, Nuclear Accident Independent Investigation Commission 2012, US Nuclear Regulatory Commission 2011). Nuclear accidents have tended to trigger organizational reform with regard to nuclear emergency response, but not on an international level. In considering this problematic ground, where might we start to develop a global approach to nuclear disaster mitigation?

Major nuclear disasters have been rare events up until now, but this might change if the nuclear sector grows in the future. With more reactors, severe accidents would have to be anticipated at greater frequency. So far, the nuclear industry has almost exclusively focused on accident prevention. While national and international disaster relief organizations have refined their response techniques over the past decades, *nuclear* emergency preparedness and response has hardly gained traction (Lakoff 2010, Redfield 2011). Public panic after the Three Mile Island accident in 1979 was arguably caused not only by 'nuclear fear', but by the public's very accurate assessment that the experts did not agree on the disaster response priorities (Public Broadcast Service 2005, Walker 2004, Weart 1988). It was not clear whether it was more important to protect the public from radiation exposure, or to save the plant, possibly even at the expense of public health. Although Three Mile Island's radiological consequences were minor compared to those of Chernobyl or Fukushima, the accident initiated a profound organizational review of the US nuclear industry. Specifically, President Jimmy Carter set up the Institute of Nuclear Power Operations in December of 1979, and tasked it with specifying safety standards in management, quality assurance, and operating practices, in an effort to *avoid* future accidents.

In the wake of the 1986 Chernobyl disaster, Soviet authorities took a different route. They created an organization, Spetsatom, and tasked it with preserving the invaluable experience gained during the disaster mitigation work, and with defining generalizable strategies about how to *respond* to a possible future nuclear emergency.[10] A few years later, however, Spetsatom was abolished and absorbed by republican 'ministries for extraordinary situations', which combined the tasks of civil defense, emergency response, and natural disaster relief. This trend toward an integrated emergency management in the former Soviet Union corresponds to developments in the United States. There, disaster relief organizations originally focused on civil defense and natural disasters. Since the end of the Cold War, these programs have gradually transformed into 'all hazard preparedness' organizations (Roberts 2010). Could such national disaster management organizations provide a blueprint for the international context? How would a nuclear emergency response group be different from 'all hazards' teams? How could we imagine dealing with unique nuclear disasters in a structured fashion?

CHALLENGES AHEAD

To move forward with maximum efficiency, an international nuclear response group needs to operationalize relevant experience from international disaster relief organizations. Its creators must learn from organizations set up to secure nuclear weapons and associated facilities. They need to analyze carefully the available experience with international governance—be that in various UN organizations or in groups pioneering international policy making such as, for example, the Intergovernmental Panel on Climate Change. Historical records on civil defense organizations should enter the debate as well, along with thorough studies of the human experience with previous nuclear accidents.[11] While direct comparisons between nuclear disasters are always problematic, there are similarities and differences worth exploring.[12] The limited experience we have with nuclear disasters suggests that in addition to thorough emergency planning, creative action has proven crucial during the aftermath of Three Mile Island and Chernobyl, and during the ongoing events at Fukushima Daiichi. Accident reports often focus on the technical measures taken to contain a problem, and frequently the personal courage and managerial ingenuity required from people on site is not mentioned or is downplayed—unless, that is, an unorthodox suggestion results in additional problems. Thus, one valuable strategy to be further developed in the nuclear industry is improvisation.[13]

Anthropologists who have studied nuclear workplaces consistently find that the 'culture of control' (that is, attempts to regulate every last action of the operating staff) is too rigid to account for all imaginable situations. Worse yet, the ideal of absolute control ignores that individuals operating high-risk technologies often have to work around unexpected events, and that flexibility is thus part of their job (Parr 2006, Perin 1998, 2005). But executives in the nuclear industry—in part driven by liability concerns—still consider it far more acceptable to introduce more rules than to systematically analyze when, why, and by whom rules might be broken. Rule-bending and judgment calls are already an integral part of many a technician's tasks at a nuclear plant; it would appear to be in the interest of overall nuclear safety to log and learn from these incidents, rather than conceal them.

Improvisation is the flip side to control and regulation, which draw on subject expertise and emphasize the execution of rules. During an emergency, improvisation typically builds on site-specific expertise and previously established informal mechanisms; it only thrives where such local knowledge, personal experience, and creative solutions are valued and acceptable contributions to a common goal (Karpan 2005, Parr 2006, Perin 1998, 2005, Roberts 1990, Weick 1987). However, it is something that nuclear organizations have not been eager to acknowledge: the uncertainties—and the stakes—are high. When considering an effective international nuclear emergency response group, improvisation should be seriously deliberated. It will be of paramount importance not just what these nuclear emergency response experts know, but also what they have experienced and how well they can lead and cooperate. There are pockets of such contextual, experience-based, 'creative' expertise, for example, embodied by 'nuclear veterans' who witnessed and handled mishaps during the early atomic age.

Even if such kinds of expertise could be successfully integrated into a nuclear emergency response team, however, the issue of internationalization remains. A nuclear emergency response group would have to deal not only with a variety of reactor designs, but also different design generations, plant layouts, and socially and culturally unique systems of operation. Finally, nuclear emergency responders would have to consider the difference in professional training that nuclear workers receive in different national contexts. Cooperation with on-site personnel, general emergency responders, and other decision makers will inevitably also run into the often underestimated problem of linguistic nuances: IAEA documents are usually translated into at least a couple of other languages, but in an emergency, time—and instant understanding—are of the essence.

There are other challenges international nuclear emergency responders would face. An operator like TEPCO, for example, might see commercial secrets jeopardized during an emergency, which might prompt it to withhold information from an international response group.[14] Also, concerns over national security and over the dual-use nature of many nuclear materials and processes would likely interfere with responders' tasks. In both cases, the transparency essential for successful emergency response operations conflicts with imperatives for confidentiality—perhaps more so than in a nonnuclear disaster (Cirincione 2007).

Another key problem for an international nuclear emergency response team is connected to recruitment, retention, and the transferability of expertise. Who would join such a group, or how would members be recruited, and what would they need to know? How would relevant knowledge be preserved, experience 'archived', improvisation made tangible, and all of this passed on to others? The training of nuclear emergency responders would also need to prepare people to be on hair-trigger alert for rare events; in other words, it would have to match a lack of relevant work routines with instantaneous availability. Technical knowledge, practice at various national simulators, and live exercises would likely have to be complemented by relevant historical and cultural training, as well as by schooling designed to develop creative thinking and resourceful action (Crossan 1998). To create such novel curricula would require intense interdisciplinary cooperation, and it is by no means clear who would ultimately pay for such projects. Associations with existing organizations could likely provide some of the necessary expertise and infrastructure.

EXISTING CANDIDATES FOR NUCLEAR EMERGENCY RESPONSE

The idea of international cooperation after a nuclear disaster is not new. In 2006, the International Atomic Energy Agency created the Response and Assistance Network (RANET), a database of national capabilities to provide 'specialized assistance by appropriately trained, equipped and qualified personnel with the ability to respond in a timely and effective manner to nuclear or radiological incidents and emergencies' (IAEA 2010: 15). On March 15, 2011, the Japanese government did request IAEA assistance, but it is not clear whether the agency subsequently involved member states through RANET.[15]

Set up in 1957 in Vienna, Austria, the IAEA was charged with promoting peaceful uses of atomic energy, with facilitating information exchange, and also with preventing the spread of nuclear weapons. This convoluted institutional agenda is in place to this day (Scheinman 1987). After the Chernobyl disaster, the agency generated two landmark conventions addressing international nuclear emergency response, the *Convention on Early Notification of a Nuclear Accident* (CENNA), and the *Convention on Assistance in the Case of a Nuclear Accident or Radiological Emergency* (CANARE) (IAEA 1986b, 1986a). The latter document set out, for the first time, a framework for international cooperation, with the IAEA providing experts and equipment to facilitate prompt assistance after a nuclear disaster.

Given its broad constituency and its multifaceted emergency response activities, the IAEA would appear to be the single most competent candidate among existing organizations to coordinate an international nuclear emergency response team. The Agency sets international standards for safety in nuclear plants and manages a dedicated Incident and Emergency Centre. In September 2011, the IAEA adopted a post-Fukushima 'Action Plan' designed to strengthen the existing nuclear safety framework by creating or expanding national rapid response teams and by encouraging member states to utilize existing assistance networks (IAEA 2011).

Despite its indisputable expertise, however, the IAEA continues to struggle with a series of impediments. For one, the agency is underresourced;

its budget for safety and security combined amounts to a fraction of other international agencies' resources (Bunn and Malin 2009: 185). In addition, the agency's mandate is complicated by its dual mission of promoting nuclear applications *and* monitoring compliance with the Nuclear Non-Proliferation Treaty. Furthermore, the IAEA has no executive authority: its safety standards are recommendations only, and its nuclear emergency response program resembles what the sociologist Lee Clarke calls 'fantasy documents'. These plans are ineffective for guiding action during an actual emergency; their main function is to assert to others 'that the uncontrollable can be controlled' (Clarke 1999: 16). Worse yet, during the developing disaster at Fukushima Daiichi, many perceived the IAEA as acting sluggishly, and as being in cahoots with the industry that some blamed for the catastrophe in the first place (Brumfiel 2011, Kurczy 2011). In the light of these persistent problems and its ambiguous public image, the IAEA would probably be wise to support the creation of an independent international nuclear disaster response team, rather than taking on the task itself.

Another candidate is the World Association of Nuclear Operators (WANO), an international organization that devotes its activities to nuclear safety. Founded in the late 1980s in response to the Chernobyl disaster WANO operates at the level of nuclear power plant operators. The Association walks a tightrope between encouraging open exchange of information among its members, and keeping plant-specific reports confidential. Similar to the IAEA, membership is voluntary, and WANO has no executive mandate. The Association's programs include peer reviews of nuclear plants, the exchange of operating experience, technical support, and professional development. In 1999, WANO signed a formal Memorandum of Understanding with the IAEA about cooperating on the set-up and management of incident reporting and analysis systems.[16] A persistent problem plaguing these systems, however, has been the underreporting of critical incidents, thus the lack of learning from them, and the resulting repetition of avoidable problems.

In April 2011, following Fukushima, WANO set up a commission to plan a more effective mitigation strategy for a future severe nuclear accident. The commission's recommendations reflected the shift from accident prevention only, to prevention and mitigation, and included (among others) the development of a worldwide integrated event response strategy.[17] According to Brian Schimmoller, a contributing editor to the industry journal *Power Engineering*, WANO's existing international infrastructure is an asset for developing a global nuclear emergency response group (Schimmoller 2012). However, at this point neither the IAEA nor WANO have the technical capabilities needed for an international nuclear emergency response team, or the personnel resources to staff one. More importantly, neither organization has any executive authority, and they both lack public trust: the IAEA due to prior inefficiency, WANO because of the confidential nature of much of its work.[18]

WANO's international programs were modeled in large part on the activities of the Institute of Nuclear Power Operations (INPO), a US industry group established in December 1979 as a direct response to Three Mile Island. Since then, INPO's central task has been to increase operational safety at US nuclear power plants. INPO coordinates plant evaluations, provides assistance with specific technical and management issues, and offers training for nuclear power professionals. Importantly, INPO also engages in events analysis and information exchange among the many private utilities operating US nuclear plants. INPO has deliberately remained off the public radar, justifying this lack of transparency with the need to facilitate harsh peer criticism among nuclear industry representatives. This strategy has earned the organization significant authority among its members, but also public criticism (Rees 1994).

In early May 2011, against the backdrop of the ongoing crisis at Fukushima, INPO's CEO, James Ellis, stepped out of the shadows to propose an international organization that would provide emergency response during accidents at nuclear energy facilities. Ellis envisioned 'a robust, highly capable response team with pre-staged equipment interoperable both domestically and internationally'.[19] This proposal was remarkable for three reasons.

First, neither Ellis, nor the organization he represents, are exactly household names. Ellis, a retired US Navy admiral, was appointed president of INPO in 2005. Under his leadership, INPO has increasingly appeared in the public media, and is usually portrayed as the reliable guarantor of US nuclear safety in the context of what some hail as a 'nuclear renaissance'.[20] But to take a proactive stand *publicly* was a clear departure from INPO's standard procedure, which emphasizes confidentiality over transparency.

Second, it was unexpected that a proposal for an *international* emergency response team would come from an organization whose tasks were confined to the United States. Ellis clearly realized that a nuclear disaster response team would face tremendous challenges at the international level. He emphasized it would be necessary 'to find the sweet spot between national sovereignty and international accountability'.[21] The latter had already become an issue in the wake of the 1986 Chernobyl disaster, when liability for transboundary contamination from a commercial nuclear facility concerned legal scholars (Malone 1987). National sovereignty, on the other hand, would presumably be affected to allow an international nuclear response group to act with maximum speed and efficiency during nuclear emergencies. Similar in some ways to the deployment of SWAT teams, elite tactical units deployed by some law enforcement agencies, a nuclear emergency group would dodge typical bureaucratic processes and override normal chains of command.

Third, the real surprise was that INPO would propose a shift from its 30-year outlook from an emphasis on *avoiding* the risk of nuclear accidents, to acknowledging the importance of an effective *response* to such events. Intentionally or not, by calling for a nuclear emergency response group (an *international* group no less), INPO dealt a direct blow to the idea that the nuclear industry could be kept accident-free in the future. The proposal thus marks an important shift of attitude within the nuclear industry itself and deserves the attention of both nuclear critics and pronuclear communities.

While the feasibility of an *international* nuclear emergency response team, as proposed by INPO and others, is still being debated, a small *national* initiative was recently set up in the United States that signals some contribution to what an international response might also feature. In June 2011, INPO, the Nuclear Energy Institute (NEI, a policy-oriented industry organization), and the Palo Alto based Electric Power Research Institute (EPRI) co-published a short position paper entitled *The Way Forward*, which outlined an integrated emergency response approach. The paper articulated eight strategic goals related to maintaining excellence (and morale) among nuclear plant workers: ensuring continued core cooling, containment integrity, and spent fuel cooling during emergencies; updating margins for protection from external events; integrating accident management guidelines with security response strategies and external event response plans; and providing clear procedures for monitoring accidental releases and accurate, timely information to the public.

In addition, the paper listed six guiding principles aimed at improving response effectiveness. These rather vague principles call for the development of diverse, flexible, and performance-based equipment and procedures for beyond-design events that also account for unique site characteristics, for strengthening cooperation with federal regulators and nonnuclear emergency response organizations, as well as for more authentic communication strategies. And finally, *The Way Forward* identified a long list of key stakeholders, including the general public, industry, regulators, emergency responders, policymakers, and opinion leaders (NEI, INPO, and EPRI 2011: 4–5). Notably, the proposal did not acknowledge that priorities are likely to differ quite significantly among these stakeholder groups.

Such strategic goals and principles provide ideas for an international response approach, and will no doubt help in suggesting mechanisms for effective international cooperation in nuclear disaster mitigation. The downside, however, is that this proposal relies exclusively on industry and private capital, and does not consider systematically what kind of unique and novel expertise would need mobilizing to make an international nuclear emergency response group succeed. Instead, *The Way Forward* is embedded in a technocratic rationality that seeks an effective 'technical fix' for reducing the risk of a nuclear disaster to manageable proportions. That misses the less tangible social expertise and improvisational skills inevitably involved in any successful disaster response.

CONCLUSION

The worst nuclear disaster in 25 years has already caused immeasurable trauma, both in Japan and around the world. In the wake of Fukushima, existing nuclear safety standards are being reviewed and upgraded, and some countries have embarked on profound policy changes (e.g. see Butler et al, and Falk, in this book). Emergency preparedness in general and preparedness for a nuclear emergency in particular, are undergoing fundamental reassessments. Within the nuclear industry, an almost exclusive emphasis on accident avoidance has given way to a new strategy of accident preparedness and response. At the broader societal level, questions are being raised about what risks are worth taking, and who should participate in these decision-making processes (see also Blowers in this book).

In this context, this chapter has focused on the emergent notion of an international nuclear emergency response group, and the challenges it will face. First, a highly skilled group of experts that could address both technical and social spheres of action, and their interrelationships, would need to be bankrolled. Given the limited budgets currently allotted to nuclear safety, this could pose a major obstacle. Second, familiarity with different reactor designs, work routines, and organizational cultures would need to be created and nurtured—even if the level of standardization in the nuclear energy sector continues to rise internationally. In this area, in particular, the cooperation of worldwide nuclear industry organizations will be crucial. And third, effective international action in a nuclear emergency will affect national sovereignty—for example, with regard to border crossing of people and equipment—at least temporarily and partially. One does not have to be a political scientist to understand how daunting the formalization of such a process would be.

If an international nuclear emergency response group is a worthwhile goal (and it certainly appears to be) we need to define realistic tasks. We need to create a credible organization—one that combines the legitimacy of a United Nations agency and the executive vigor of an industry group. Perhaps most importantly, we need to think hard about how these responders are recruited, what they need to know, and how they learn it. It is imperative to merge the discipline, ethics, and technical proficiencies indispensable for this task with skills that underscore flexibility and improvisation. To accomplish this, an interdisciplinary integration of technical and social science competencies will be needed. Scholars in science, technology and society studies no doubt have much to offer—but it remains to be seen whether the 'nuclear village' is ready to collaborate in the international governance of nuclear emergencies.

NOTES

I thank Richard Hindmarsh, Alireza Haghighat, Dieter Sperl, Philip Egert, and—as always—Irina and Iouli Andreev for stimulating comments on earlier drafts of this chapter.

1. In this chapter 'emergency response group' refers to a dedicated team of people who could rapidly dispatch to the site of a nuclear disaster.
2. I will use the shorthand 'Fukushima' unless I refer to plant-specific problems at Fukushima Daiichi.
3. It will be more difficult to blame technical and managerial incompetence for Fukushima Daiichi than it was for Chernobyl (for an initial international

assessment, that has since been significantly modified, see INSAG-1 1986; for a short overview of revisions of these initial reports see Schmid 2011c).

4. For the following summary, I rely on official reports such as Government of Japan 2011, INPO 2011, Investigation Committee 2011, US Nuclear Regulatory Commission 2011, and various media reports.
5. For a rebuttal by General Electric Co. see, for example, "Setting the Record Straight on Mark I Containment History" *GE Reports*, March 18, 2011, accessed June 12, 2012, http://www.gereports.com/setting-the-record-straight-on-mark-i-containment-history See also US Nuclear Regulatory Commission 2012.
6. See, for example, Suzuki 2011, Fujigaki and Tsukahara 2011. See also the following journal special issues on the topic: *Ethics, Policy & Environment* 14, no. 3 (2011); *Environmental History* 17, no. 2 (April 2012); *Bulletin of the Atomics Scientists,* May/June 2012.
7. Group of Eight summit, May 26–27, 2011, Deauville, France, http://www.g20-g8.com/g8-g20/g8/english/the-2011-summit/declarations-and-reports/appendices/report-of-the-nuclear-safety-and-security-group.1355.html; see also "France Calls for Global Nuclear Safety Standards," Voice of America, March 31, 2011, http://www.voanews.com/english/news/europe/Sarkozy-Calls-For-Global-Nuclear-Safety-Standard-By-Years-End-118977029.html. Both accessed June 12, 2012.
8. "EurActiv G8 Summit Urges Stringent Nuclear Safety Rules," May 27, 2011, accessed June 12, 2012, http://www.euractiv.com/energy/g8-summit-urges-stringent-nuclear-safety-rules-news-505185.
9. The discussion over the impacts of radiation fallout on ecosystems is not new (Paine 1992, Yablokov et al. 2009, Wynne 2010) but has received renewed attention after Fukushima. Likewise, the debate over human dose thresholds has flared up again after Fukushima but has accompanied the development of nuclear energy since the 1940s (see, e.g. the famous 'Baby tooth survey' [Reiss 1961]; more recent reports such as Mangano and Sherman [2011]; and the last in a series of EPA reports on health effects of radiation: BEIR Committee [2006]).
10. Personal communication with the former scientific director of Spetsatom, Iouli B. Andreev, December 2011, January 2012, Vienna, Austria.
11. For example, organizers of the Chernobyl disaster mitigation work are still alive but they are often ailing and advancing in age. Oral histories with these specialists may be a first step but cannot suffice. Stanford's Hoover Institution has begun to collect narrative interviews with leading Chernobyl disaster mitigation workers.
12. STS research has shown the value of comparing economic structures and regulatory cultures when assessing the sometimes stark differences between how emerging technologies fare in different countries (e.g. Jasanoff 2005, Nelkin and Pollak 1981, Parthasarathy 2007).
13. A large body of management literature addresses the value of improvisation for organizations (e.g. Kamoche et al. 2002). To my knowledge, this work has not yet been utilized for the study of organizations in high-risk industries.
14. Trade secrets could relate to design modification, cooperative agreements, or routines of handling specific tasks and processes.
15. "Fukushima Nuclear Accident Update (15 March 2011, 20: 35 UTC)," accessed June 12, 2012, http://www.iaea.org/newscenter/news/2011/fukushima150311.html.
16. WANO, "History," accessed June 12, 2012, http://www.wano.info/about-us/history/.

17. WANO, "WANO Members Unanimously Approve New Commitments to Nuclear Safety," October 25, 2011, accessed June 12, 2012, http://www.wano.info/2011/10/wano-biennial-general-meeting-press-release/.
18. "Cooperation in the Post-Fukushima Era," *World Nuclear News*, 21 June 2011, accessed July 28, 2012, http://www.world-nuclear-news.org/RS_Cooperation_for_the_post_Fukushima_era_2106111.html; Bunn 2011; Braithwaite and Drahos 2000: 297–321.
19. Nuclear Energy Institute, "INPO Chief Proposes Global Nuclear Response Group," May 12, 2011, accessed June 12, 2012, http://www.nei.org/newsandevents/conferencesandmeetings/nea/nuclear-energy-assembly-2011-news-coverage/inpo-chief-proposes-global-nuclear-response-group/.
20. For example, Lauvergeon 2009; World Nuclear Association, "The Nuclear Renaissance," accessed June 12, 2012, http://www.world-nuclear.org/info/inf104.html. For a skeptical view see Feiveson 2007, 2009.
21. See note 18.

REFERENCES

Aldrich, D. "Future Fission: Why Japan Won't Abandon Nuclear Power." *Global Asia* 6, no. 2 (2011): 63–67.

Beck, U. *Risk Society: Towards a New Modernity*. London: Sage, 1992.

BEIR Committee. *Health Risks from Exposure to Low Levels of Ionizing Radiation (BEIR-VII)*. Washington, DC: National Academic Press, 2006.

Bosk, M. "Fukushima: Mark 1 Nuclear Reactor Design Caused GE Scientist to Quit in Protest." *ABC News*, March 15, 2011.

Braithwaite, J., and P. Drahos. *Global Business Regulation*. Cambridge: Cambridge University Press, 2000.

Brumfiel, G. "Nuclear Agency Faces Reform Calls: International Atomic Energy Agency's Remit under Scrutiny" *Nature*, April 26, 2011.

Bunn, M. "Mostly Getting Nuclear Safety at the IAEA—but Missing Nuclear Security." *Power & Policy*, June 21, 2011, accessed July 28, 2012. http://www.powerandpolicy.com/2011/06/21/mostly-getting-nuclear-safety-at-the-iaea-%E2%80%93-but-missing-nuclear-security/.

Bunn, M., and M. Malin. "Enabling a Nuclear Revival—and Managing its Risks." *Innovations* 4, no. 4 (2009): 173–191.

Butler, C., K. Parkhill, and N. Pidgeon. "Nuclear Power after Japan: The Social Dimensions." *Environment* 63, no. 6 (2011): 3–14.

Chan, C., and Y. Chen. "A Fukushima-Like Nuclear Crisis in Taiwan or a Non-nuclear Taiwan?" *East Asian Science, Technology and Society: An International Journal* 5 (2011): 403–407.

Cirincione, J. *Bomb Scare: The History and Future of Nuclear Weapons*. New York: Columbia University Press, 2007.

Clarke, L. *Mission Improbable: Using Fantasy Documents to Tame Disaster*. Chicago and London: Chicago University Press, 1999.

Collins, H., and R. Evans. "The Third Wave of Science Studies: Studies of Expertise and Experience." *Social Studies of Science* 32, no. 2 (2002): 235–296.

Crossan, M. "Improvisation in Action." *Organization Science* 9, no. 5 (1998): 593–599.

European Nuclear Safety Regulators Group. "Peer Review Report: Stress Tests Performed on European Nuclear Power Plants." 2012, accessed June 12, 2012. http://www.ensreg.eu.

Feiveson, H. "Faux Renaissance: Global Warming, Radioactive Waste Disposal, and the Nuclear Future." *Arms Control Today,* May 1, 2007.
Feiveson, H. "A Skeptic's View of Nuclear Energy." *Daedalus* 138, no. 4 (2009): 60–70.
Fujigaki, Y., and T. Tsukahara. "STS Implications of Japan's 3/11 Crisis." *East Asian Science, Technology and Society: An International Journal* 5 (2011): 381–394.
Government of Japan, Nuclear Emergency Response Headquarters. "Report of the Japanese Government to the IAEA Ministerial Conference on Nuclear Safety: The Accident at TEPCO's Fukushima Nuclear Power Stations." [Tokyo:] June 2011.
Guston, D., and D. Sarewitz. "Real-Time Technology Assessment." *Technology in Society* 24, nos. 1–2 (2002): 93–109.
Hong, S. "Where Is the Nuclear Nation Going? Hopes and Fears over Nuclear Energy in South Korea after the Fukushima Disaster." *East Asian Science, Technology and Society: An International Journal* 5 (2011): 409–415.
INSAG-1. "Summary Report on the Post-Accident Review Meeting on the Chernobyl Accident." Safety Series No. 75-INSAG-1. Vienna: International Atomic Energy Agency, 1986.
Institute of Nuclear Power Operations. *Special Report on the Nuclear Accident at the Fukushima Daiichi Nuclear Power Station.* Atlanta, GA, 2011.
IAEA (International Atomic Energy Agency). *Treaty on the Nonproliferation of Nuclear Weapons.* INFCIRC/140, April 22, 1970.
IAEA (International Atomic Energy Agency). *Convention on Assistance in the Case of a Nuclear Accident or Radiological Emergency* (CANARE). IAEA-INFCIRC-336. Vienna: IAEA, November 18, 1986a.
IAEA (International Atomic Energy Agency). *Convention on Early Notification of a Nuclear Accident* (CENNA). INFCIRC/335 Vienna: IAEA, November 18, 1986b.
IAEA (International Atomic Energy Agency). *IAEA Response and Assistance Network, Incident and Emergency Centre.* Vienna: IAEA, 2010.
IAEA (International Atomic Energy Agency). *Draft IAEA Action Plan on Nuclear Safety.* GOV/2011/59-GC(55)/14, September 5, 2011.
Investigation Committee on the Accident at the Fukushima Nuclear Power Stations of Tokyo Electric Power Company. *Interim Report.* 2011, accessed June 12, 2012. http://icanps.go.jp/eng/interim-report.html.
Jahn, D., and S. Korolczuk. "German Exceptionalism: The End of Nuclear Energy in Germany!" *Environmental Politics* 21, no. 1 (2012): 159–164.
Jasanoff, S. (ed). *Learning from Disaster: Risk Management after Bhopal.* Philadelphia: University of Pennsylvania Press, 1994.
Jasanoff, S. *Designs on Nature: Science and Democracy in Europe and the United States.* Princeton, NJ: Princeton University Press, 2005.
Jiang, Z. "Fukushima Nuclear Accident Implications for the Nuclear Power Development in China." *Advanced Materials Research* 347–353 (2012): 3810–3814.
Kamoche, K., M. Cunha, and J. Cunha, eds. *Organizational Improvisation.* New York: Routledge, 2002.
Karpan, N. *Chernobyl: 'Mest' mirnogo atoma.* Kiev: ChP Kantri Laif, 2005.
Kingdon, J. *Agendas, Alternatives, and Public Policies.* New York: Longman, 2003.
Kleinman, D. (ed). *Science, Technology, and Democracy.* Albany: State University of New York Press, 2000.
Kurczy, S. "Japan Nuclear Crisis Sparks Calls for IAEA Reform." *Christian Science Monitor,* March 17, 2011.
Lakoff, A. (ed). *Disaster and the Politics of Intervention.* New York: Columbia University Press, 2010.
Lauvergeon, A. "The Nuclear Renaissance: An Opportunity to Enhance the Culture of Nonproliferation." *Daedalus* 138, no. 4 (2009): 91–99.

Lente, D. van (ed). *The Nuclear Age in Popular Media: A Transnational History, 1945–1965*. New York: Palgrave Macmillan, 2012.

Lupton, D. *Risk*. London & New York: Routledge, 1999.

Malone, L. "The Chernobyl Accident: A Case Study in International Law Regulating State Responsibility for Transboundary Nuclear Pollution." *Journal of Environmental Law* 12 (1987): 203–241.

Mangano, J., and J. Sherman. "Elevated in Vivo Strontium-90 from Nuclear Weapons Test Fallout among Cancer Decedents: A Case-Control Study of Deciduous Teeth." *International Journal of Health Sciences* 41, no. 1 (2011): 137–158.

Meshkati, N. "We Must Cooperate on Nuclear Safety." *New York Times*, September 22, 2011.

Morita, H. *Rescuing Victims and Rescuing TEPCO: A Legal and Political Analysis of the TEPCO Bailout*. March 21, 2012, accessed June 12, 2012. http://ssrn.com/abstract=2026868.

NEI, INPO, and EPRI. *The Way Forward. U.S. Industry Leadership in Response to Events at the Fukushima Daiichi Nuclear Power Plant*. Nuclear Energy Institute, Institute of Nuclear Power Operations, and Electric Power Research Institute. June 8, 2011, accessed June 12, 2012. http://www.nei.org/filefolder/Way_Forward_web.pdf.

Nelkin, D. (ed). *Controversy: Politics of Technical Decisions*. 3rd ed. Newbury Park: Sage, 1992.

Nelkin, D., and M. Pollak. *The Atom Besieged: Extraparliamentary Dissent in France and Germany*. Cambridge, MA, and London: MIT Press, 1981.

Nuclear Accident Independent Investigation Commission. *The Official Report of the Fukushima Accident Independent Investigation Commission*. [Tokyo:] National Diet of Japan, 2012.

Otway, H., and B. Wynne. "Risk Communication: Paradigm and Paradox." *Risk Analysis* 9 (1989): 141–145.

Paine, R. "Chernobyl Reaches Norway: The Accident, Science, and the Threat to Cultural Knowledge." *Public Understanding of Science* 1 (1992): 261–280.

Parr, J. "A Working Knowledge of the Insensible? Radiation Protection in Nuclear Generating Stations, 1962–1992." *Comparative Studies in Society and History* 48, no. 4 (2006): 820–851.

Parthasarathy, S. *Building Genetic Medicine: Breast Cancer, Technology, and the Comparative Politics of Health Care*. Cambridge, MA: MIT Press, 2007.

Perin, C. "Operating as Experimenting: Synthesizing Engineering and Scientific Values in Nuclear Power Production." *Science, Technology & Human Values* 23, no. 1 (1998): 98–128.

Perin, C. *Shouldering Risks: The Culture of Control in the Nuclear Power Industry*. Princeton, NJ: Princeton University Press, 2005.

Perrow, C. *Normal Accidents: Living with High-Risk Technologies*. New York and London: Basic Books, 1984.

Perrow, C. *The Next Catastrophe: Reducing Our Vulnerabilities to Natural, Industrial, and Terrorist Disasters*. Princeton, NJ: Princeton University Press, 2007.

Pritchard, S. "An Envirotechnical Disaster: Nature, Technology, and Politics at Fukushima." *Environmental History* 17 (2012): 219–243.

Public Broadcast Service. *The American Experience: Meltdown at Three Mile Island*. VHS. 2005

Redfield, P. "Cleaning Up the Cold War: Global Humanitarianism and the Infrastructure of Crisis Response." In *Entangled Geographies: Empire Technopolitics in the Global Cold War*, edited by G. Hecht, 267–291. Cambridge, MA: MIT Press, 2011.

Rees, J. *Hostages of Each Other: The Transformation of Nuclear Safety since Three Mile Island*. Chicago and London: Chicago University Press, 1994.

Reiss, L. "Strontium-90 Absorption by Deciduous Teeth." *Science* 134 (1961): 1669–1673.

Roberts, K. "Some Characteristics of One Type of High Reliability Organization." *Organization Science* 1, no. 2 (1990): 160–176.

Roberts, P. "Private Choices, Public Harms: The Evolution of National Disaster Organizations in the United States." In *Disaster and the Politics of Intervention*, edited by A. Lakoff, 42–69. New York: Columbia University Press, 2010.

Sagan, S. *The Limits of Safety: Organizations, Accidents, and Nuclear Weapons.* Princeton, NJ: Princeton University Press, 1993.

Scheinman, L. *The International Atomic Energy Agency and World Nuclear Order.* Baltimore: Johns Hopkins University Press, 1987.

Schimmoller, B. "Will the New WANO Have Enough Teeth?" *Power Engineering*, January 1, 2012.

Schmid, S. "Both Better and Worse than Chernobyl." *London Review of Books Blog*, March 17, 2011a, accessed June 12, 2012. http://www.lrb.co.uk/blog/2011/03/17/sonja-schmid/both-better-and-worse-than-chernobyl/.

Schmid, S. "The Unbearable Ambiguity of Knowing: Making Sense of Fukushima." *Bulletin of the Atomic Scientists* (web edition). April 11, 2011b, accessed June 12, 2012. http://www.thebulletin.org/web-edition/features/the-unbearable-ambiguity-of-knowing-making-sense-of-fukushima.

Schmid, S. "When Safe Enough Is Not Good Enough: Organizing Safety at Chernobyl." *Bulletin of the Atomic Scientists* 67, no. 2 (2011c): 19–29.

Sclove, R. *Reinventing Technology Assessment: A 21st Century Model.* Washington, DC: Science and Technology Innovation Program, Woodrow Wilson International Center for Scholars, April 2010, accessed June 12, 2012. http://wilsoncenter.org/techassessment.

Sebeok, T. *Communication Measures to Bridge Ten Millennia*. Columbus, OH: Battelle Memorial Institute, Office of Nuclear Waste Isolation, 1984.

Shrader-Frechette, K. "Fukushima, Flawed Epistemology, and Black-Swan Events." *Ethics, Policy and Environment* 14, no. 3 (2011): 267–272.

Suzuki, T. "Deconstructing the Zero-Risk Mindset: The Lessons and Future Responsibilities for a Post-Fukushima Nuclear Japan." *Bulletin of the Atomic Scientists* 67, no. 5 (2011): 9–18.

Szerszynski, B. "Risk and Trust: The Performative Dimension." *Environmental Values* 8, no. 2 (1999): 239–252.

Takubo, M. "Nuclear or Not? The Complex and Uncertain Politics of Japan's Post-Fukushima Energy Policy." *Bulletin of the Atomic Scientists* 67, no. 5 (2011): 19–26.

Thomas, S. "What Will the Fukushima Disaster Change?" *Energy Policy* 45 (2012): 12–17.

US Nuclear Regulatory Commission. "Recommendations for Enhancing Reactor Safety in the 21st Century: The Near-Term Task Force Review of Insights from the Fukushima Dai-ichi Accident." [Washington, DC] 2011.

US Nuclear Regulatory Commission. "Resolution of Generic Safety Issues: Issue 157: Containment Performance (Rev. 1)." NUREG-0933, Main Report with Supplements 1–34. Last updated March 29, 2012.

Vaughan, D. *The Challenger Launch Decision: Risky Technology, Culture, and Deviance at NASA*. Chicago: University of Chicago Press, 1996.

Vaughan, D. "Organizational rituals of risk and error." In *Organizational Encounters with Risk,* edited by B. Hutter and M. Power, 33–66. Cambridge: Cambridge University Press, 2005.

Wald, M., and J. Berger, "Regulator Says Fuel Pools at U.S. Reactors Are Ready for Emergencies." *New York Times,* March 20, 2011.

Walker, J. *Containing the Atom: Nuclear Regulation in a Changing Environment, 1963–1971*. Berkeley: University of California Press, 1992.
Walker, J. *Three Mile Island: A Nuclear Crisis in Historical Perspective*. Berkeley: University of California Press, 2004.
Weart, S. *Nuclear Fear: A History of Images*. Cambridge, MA and London: Harvard University Press, 1988.
Weick, K. "Organizational Culture as a Source of High Reliability." *California Management Review* 29, no. 2 (1987): 112–127.
Weinberg, A. "Social Institutions and Nuclear Energy." *Science* 177, no. 4043 (1972): 27–34.
Wynne, B. "May the Sheep Safely Graze? A Reflexive View of the Lay-Expert Divide." In *Risk, Environment, and Modernity*, edited by S. Lash, B. Szerszynski, and B. Wynne, 44–83. London: Sage, 1996.
Wynne, B. *Rationality and Ritual: Participation and Exclusion in Nuclear Decision-Making*. London and New York: Routledge, 2010.
Yablokov, A., V. Yablokov, V. Nesterenko, and A. Nesterenko. *Chernobyl: Consequences of the Catastrophe for People and the Environment*. Boston, MA: Blackwell: Annals of the New York, Academy of Sciences, 2009.
Zeller Jr., T., "Experts Had Long Criticized Potential Weakness in Design of Stricken Reactor." *New York Times*, March 15, 2011.
Zhou, Y., C. Rengifo, P. Chen, and J. Hinze. "Is China Ready for Its Nuclear Expansion?" *Energy Policy* 39, no. 2 (2011): 771–781.

12 Fallout from Fukushima Daiichi

An Endnote

Richard Hindmarsh

The contributors to this book—*Nuclear Disaster at Fukushima Daiichi*—have penned a variety of well-documented and insightful accounts of the disaster. Informed by the field of science, technology and society (STS) studies these accounts serve to enhance our understanding of how and why the Fukushima Daiichi nuclear disaster occurred, especially in 'socio-political, alternatively, sociotechnical', context. They have well identified and addressed key social, political and environmental impacts and issues that the disaster raised and highlighted; and what might be done by way of social and policy learning to address the implications of those impacts and issues.

In this endeavor the contributors identified and addressed six key themes that characterized the issues, concerns and debates about Fukushima Daiichi, with each contributor investigating one or more themes in their chapters. These themes were the social shaping and subsequent compromise of nuclear power safety; the flawed concentrated siting of nuclear power plants (or stations) and reactors and accompanying inadequate public participation and community engagement about such siting; the failure of official risk communication to citizens involving radiation data and other relevant information postdisaster; the actual nature of the disaster—that is, to what extent the social realm could be seen as causative alongside the natural; the future of nuclear energy and the alternatives; and the failings of emergency responses to manage adequately the disaster as it was unfolding and in its immediate aftermath.

All of these themes pose profound implications for Japan's nuclear energy program and energy security, as well as sustainability futures, with many also posed for the global nuclear industry, in a number of important social, environmental, policy and energy contexts. All were found to inform the disaster as a social, technological and natural disaster, that is, as a *chronic technological and natural disaster* (as Chapter 1 outlines in detail), alternatively as an 'envirotechnical' one. Shaping the social side of the disaster was a *policy in action* terrain of human actions, policies and decisions or lack thereof. This discursive terrain well reflected the strategic social context of key nuclear power development areas of safety regulation, planning and power plant siting, public participation, and rather belatedly, disaster management. Following the contributors' investigations and analysis, all of these areas were found to be in much need of enhanced and more responsible science, technology and environmental governance, according to notions of good governance principles, for example, of openness, (civic) participation, accountability, effectiveness and coherence for policy learning and change.

In arriving at these findings, the critique was sharp, informative and insightful. It makes a significant if not seminal contribution to a burgeoning literature on the social, political and policy context of the Fukushima disaster both in Japan and globally. Japan's nuclear power governance system was found overwhemingly to have long discounted known safety problems of nuclear power both in operation and siting, made worse in a siting landscape of high seismic instability and hazard. In favor of limited technocratic planning approaches based on technical data and narrow, expert pro-nuclear build perceptions, the need for integrated or synergistic planning approaches that bridge socially constructed expert/lay and social/technical divides or boundaries were overlooked. Such integrated approaches are now increasingly being adopted worldwide in environmental planning and natural resource management (and many other areas) to best manage uncertainty and complexity to avoid ill-conceived planning outcomes and critical system failures. Instead, under Japan's technically biased planning and regulatory approach (which, of course, reflects an approach that still lingers worldwide), relational or tightly coupled sociotechnical system components were often seen largely as easy-to-manage and relatively isolated or separate parts. This parts approach coupled to a complacent view of the technical superiority over the social and over nature, made this approach highly susceptible to compromise and failure. Such failure was furthered intimately by way of a close-knit network of, or cozy relationship between, powerful public officials and industry interests intent on facilitating nuclear build as much and as fast as possible.

Weak regulation—seemingly reflecting increasing agency capture, or dependence, of the regulator to or on the nuclear industry—was the order of the day to progress nuclear development. Failure struck many times over the decades with a string of nuclear incidents in Japan, which gave clear warning of the problems of this reductionist and technocratic approach to the safety and siting of nuclear power. At the end, with the Fukushima Daiichi disaster, these problems could no longer be denied or shunted aside with this unprecedented event of a three reactor meltdown, all at one site, and with the subsequent insidious threat and impact of long-term radioactive pollution.

The implication of this dramatic and tragic failure embedded in such a flawed nuclear power management approach strengthens the need for participatory and precautionary integrated development, planning and regulatory approaches for controversial technoscience. It serves as a clear lesson, even warning, worldwide to move beyond limited fragmented or so-called silo approaches to megatechnological and/or envirotechnical management. Clearly, a strong implication of the disaster is that Japan's, and broader global, nuclear power planning and regulatory systems need to catch up posthaste to integrated management styles based on principles of good governance and long-term sustainability. More broadly, given the transboundary regional and global impacts of Fukusima, a precautionary, collaborative and coherent planning approach for siting nuclear power plants internationally is needed when they are planned.

By association, another implication of the Fukushima disaster is the need to decouple 'interdependencies' or 'cozy' policy interactions between pro-nuclear government officials and regulators and the nuclear industry, which were finally acknowledged in Japan after Fukushima occurred, although it seems quite reluctantly, as compromising safety. Facing a highly volatile political situation of public dissent and distrust, the Japanese Government launched a new so-called independent regulatory agency in September 2012. Soon, however, reflecting again the 'closed' insider-governmental and expert domination of the new regulatory authority and the long endemic problem of collusion between regulation and industry, problems of some members with past collusion quickly emerged, reinforcing issues of public distrust and preferences for non-nuclear energy futures. What such complicity reinforces again—in this high-stakes, and highly sensitive and risky, megatechnological area—is the need for a highly responsible and transparent civic regulatory style to better inform nuclear development, safety and siting issues of complexity, risk, trust, civic inclusion and legitimacy, and, in turn, policy effectiveness.

Alternatively, integrated approaches address another policy implication of nuclear power development in Japan that Fukushima Daiichi well demonstrated. This implication is the dire need for *functional* sociotechnical systems involving large technological facilities, particularly megatechnological ones. For functionality and policy effectiveness and legitimacy the facility (the technical infrastructure)—here the nuclear power plant or, for that matter, any other controversial energy facility, such as the large wind farm—has to be well embedded or interconnected to its social infrastructure (the social and civic spheres). In Japan nuclear power development was largely isolated from its social infrastructure, which increased the likelihood of some sort of system *breakdown* being inevitable, as occurred in the tragic case of Fukushima. Functionality of this type thus needs effective sociotechnical interconnectedness. Social functionality for the technical at the community level has been found by a burgeoning literature to be best realized through a range of inclusive participatory approaches. These approaches feature open and full information, transparency and accountability, and enhanced citizen participation at the broad societal level, to highly collaborative community engagement enabling strong decisional influence at the local level regarding controversial facility siting posing radical change. The concept of *place-change planning* was especially advanced in this regard.

Deepening such arguments, the contributors also advanced the need for improved emergency responses *locally* in involving citizen science through the new concept and development of civic radiation maps and low-cost DIY (do-it-yourself) radiation monitoring devices and the wide use of social media for enhanced environmental and disaster management. In this context, the contribution of science citizen empowerment in reflecting *cosmopolitical citizenship* was advanced. *Globally*, a convincing case was also made for the development of a nuclear SWAT team as an international nuclear emergency response group in the somber reflection that Fukushima will likely not remain the last severe nuclear accident global society will have to face.

Turning to the problematic of nuclear energy futures and the ongoing dire lack of nuclear 'fail safe' over many decades, the view where expressed by contributors was that renewable energy poses a better option than nuclear, particularly in being radiation free, generating better public trust and confidence about energy safety and desirability, and offering greater sustainability. At the same time, however, renewable transitions—along with enhanced energy conservation and efficiency—although slowly gathering strength, will not likely occur in a significant way in the immediate and perhaps foreseeable future in many countries (with nuclear phaseouts being quite long—in the case of Japan a suggested 30 years, and with many nuclear nations maintaining a nuclear energy agenda). For example, the International Atomic Energy Agency reported in 2011 that some 60 countries were considering introducing nuclear power. But perhaps this is already changing post-Fukushima?

Another clear implication of this ongoing nuclear situation, even in a relatively short time frame, is that obviously, safety regulation and implementation need to be significantly tightened and revamped for nuclear power plants, their associated radioactive waste repositories, their facility siting, and, of course, the accompanying but limited technical regulatory styles worldwide. However, compounding the reformist task for such safety enhancement to be effective and also for the suggestion of a nuclear SWAT team, is the implication of Fukushima as a *new type of major nuclear disaster* marking the close interactivity of the social, technological and natural disaster terrains, which set it apart from Three Mile Island and Chernobyl on the natural disaster front. This interactivity, especially regarding the natural, then opens up new complex questions about adequate reactor design, safety and siting, and thus, the adequacy of emergency responses. Such complexity is furthered by the threat of escalating extreme weather events associated with climate change, for example, in addition to the hazardous seismic activities of the Pacific Basin or elsewhere. Such decision making involving hazardous megatechnologies and terrains again reinforces the need for participatory and precautionary integrated energy management and policy approaches in contexts of legitimacy and policy effectiveness.

In conclusion, the insights, findings and suggestions of the contributors to this book highlight the interactivity of all key aspects of nuclear power technically and socially in relation to the social, political and environmental issues and implications of the Fukushima Daiichi nuclear disaster. The main points suggested here, at the least, for social and policy learning of nuclear power management—both in Japan and globally—is that rapid development of good governance and integrated participatory management approaches is needed. This needs to conjoin expert and civic knowledge areas and increase legitimacy through enhanced public participation providing inclusiveness, transparency, open and full information, dialogue and accountability in decision making, which, in turn, poses increased energy policy and management effectiveness. Finally, there is the parallel need to adopt renewable energies alongside enhanced energy efficiency and conservation where possible to replace nuclear power in the interests of long-term energy safety, security and sustainability.

Such suggestions work into arguably the overarching implication of the Fukushima Daiichi nuclear disaster, as reinforced by Three Mile Island and Chernobyl and many other serious nuclear power incidents and accidents since the early 1950s. This implication is to fundamentally reshape the way in which many people, regions and countries are perceiving the *risks and hazards* of nuclear power in relation to health, safety, energy security, participation, good governance, and the environment and sustainability when considering existing and future energy pathways.

Index

CPSIA information can be obtained
at www.ICGtesting.com
Printed in the USA
JSHW011454201219
3107JS00006B/144